T0276814

Chemical Engineering: Advanced Concepts

Chemical Engineering: Advanced Concepts

Edited by **Jina Redlin**

CLANRYE
INTERNATIONAL

New Jersey

Published by Clanrye International,
55 Van Reypen Street,
Jersey City, NJ 07306, USA
www.clanryeinternational.com

Chemical Engineering: Advanced Concepts
Edited by Jina Redlin

© 2015 Clanrye International

International Standard Book Number: 978-1-63240-095-6 (Hardback)

Contents

Preface

This book has been an outcome of determined endeavour from a group of educationists in the field. The primary objective was to involve a broad spectrum of professionals from diverse cultural background involved in the field for developing new researches. The book not only targets students but also scholars pursuing higher research for further enhancement of the theoretical and practical applications of the subject.

This book discusses current changes in chemical engineering and gives a comprehensive summary of the developments occurring within this field. From the kinetic and thermodynamic molding aspect, the molecular angle is extremely significant. Hence, most of the issues discussed within this book were written keeping in mind industrial problems and posit solutions to these problems from experts around the world. The contents of the book broadly belong to two categories, process engineering and separation technology. This book will provide fundamental knowledge regarding this field to both students and professionals.

It was an honour to edit such a profound book and also a challenging task to compile and examine all the relevant data for accuracy and originality. I wish to acknowledge the efforts of the contributors for submitting such brilliant and diverse chapters in the field and for endlessly working for the completion of the book. Last, but not the least; I thank my family for being a constant source of support in all my research endeavours.

Editor

Part 1

Process Engineering

CFD Modelling of Fluidized Bed with Immersed Tubes

A.M.S. Costa, F.C. Colman, P.R. Paraiso and L.M.M. Jorge
Universidade Estadual de Maringa
Brazil

1. Introduction

Fluidized beds are widely used in combustion and chemical industries. The immersed tubes are usually used for enhancement of heat transfer or control of temperature in fluidized beds. By his turn, tubes subjected to the solid particle impact may suffer severe erosion wear. Many investigations have been devoted to erosion in tubes immersed in fluidized beds on the various influencing factors (cf. Lyczkowski and Bouillard, 2002). As pointed by Achim et al. (2002), the factors can be classified as particle characteristics, mechanical design and operating conditions.

Some previous experimental studies have focused on bubble and particle behaviors (Kobayashi et al., 2000, Ozawa et al., 2002), tube attrition, erosion or wastage (Bouillard and Lyczkowski, 1991; Lee and Wang, 1995; Fan et al., 1998; Wiman, 1994), heat transfer (Wong and Seville, 2006, Wiman and Almstedt, 1997) and gas flow regimes (Wang et. al, 2002).

Previous numerical studies were also performed using different CFD codes. Recently He et al. (2009, 2004), using the K-FIX code adapted to body fitted coordinates investigated the hydrodynamics of bubbling fluidized beds with one to four immersed tubes. The erosion rates predicted using the monolayer kinetic energy dissipation model were compared against the experimental values of Wiman (1994) for the two tube arrangement. The numerical values were three magnitudes lower than the experimental ones. Also employing a eulerian-eulerian model and the GEMINI numerical code, Gustavsson and Almstedt (2000, 1999) performed numerical computations and comparison against experimental results (Enwald et. al., 1999). As reported for those authors, fairly good qualitative agreement between the experimental and numerical erosion results were obtained, and the contributions to the erosion from the different fluid dynamics phenomena near the tube were identified.

In the present study, we revisit the phenomena of the immersed tubes in a gas fluidized bed with different immersed tube arrangements employing the eulerian-eulerian two fluid model and the MFIX code. The purposes of the numerical simulations are to compare and explore some effects not previously investigated in the above-mentioned references.

2. Two fluid and erosion model

The mathematical model is based on the assumption that the phases can be mathematically described as interpenetrating continua; the point variables are averaged over a region that is large compared with the particle spacing but much smaller than the flow domain (see Anderson, 1967). A short summary of the equations solved by the numerical code in this study are presented next. Refer to Benyahia et al. (2006) and Syamlal et al. (1993) for more detailment.

The continuity equations for the fluid and solid phase are given by :

$$\frac{\partial}{\partial t}(\varepsilon_f \rho_f) + \nabla \cdot (\varepsilon_f \rho_f \, \vec{v}_f) = 0 \tag{1}$$

$$\frac{\partial}{\partial t}(\varepsilon_s \rho_s) + \nabla \cdot (\varepsilon_s \rho_s \, \vec{v}_s) = 0 \tag{2}$$

In the previous equations ε_f, ε_s, ρ_f, ρ_s, $\vec{v}_f$ and $\vec{v}_s$ are the volumetric fraction, density and velocity field for the fluid and solids phases.

The momentum equations for the fluid and solid phases are given by:

$$\frac{\partial}{\partial t}(\varepsilon_f \rho_f \vec{v}_f) + \nabla \cdot (\varepsilon_f \rho_f \vec{v}_f \vec{v}_f) = \nabla \cdot \overline{\overline{S}}_f + \varepsilon_f \rho_f \vec{g} - \overline{I}_{fs} \tag{3}$$

$$\frac{\partial}{\partial t}(\varepsilon_s \rho_s \vec{v}_s) + \nabla \cdot (\varepsilon_s \rho_s \vec{v}_s \vec{v}_s) = \nabla \cdot \overline{\overline{S}}_s + \varepsilon_s \rho_s \vec{g} + \overline{I}_{fs} \tag{4}$$

$\overline{\overline{S}}_f$, $\overline{\overline{S}}_s$ are the stress tensors for the fluid and solid phase. It is assumed newtonian behavior for the fluid and solid phases, i.e.,

$$\overline{\overline{S}} = \left(-P + \lambda \, \nabla \cdot \vec{v}\right)\overline{\overline{I}} + 2\mu \, S_{ij} \equiv -p \, \overline{\overline{I}} + \overline{\overline{\tau}} \qquad S_{ij} = \frac{1}{2}\left[\nabla \vec{v} + (\nabla \vec{v})^T\right] - \frac{1}{3}\nabla \cdot \vec{v} \tag{5a,b}$$

In the above equation P, λ, μ are the pressure, bulk and dynamic viscosity, respectively.

In addition, the solid phase behavior is divided between a plastic regime (also named as slow shearing frictional regime) and a viscous regime (also named as rapidly shearing regime). The constitutive relations for the plastic regime are related to the soil mechanics theory. Here they are represented as:

$$p_s^p = f_1\left(\varepsilon^*, \varepsilon_f\right). \qquad \mu_s^p = f_2\left(\varepsilon^*, \varepsilon_f, \phi\right) \tag{6}$$

In the above equation ε^* is the packed bed void fraction and ϕ is the angle of internal friction.

A detailing of functions f_1 to f_4 and f_9 can be obtained in Benyahia (2008).

On the other hand, the viscous regime behavior for the solid phase is ruled by two gas kinetic theory related parameters (e, Θ).

$$p_s^v = f_3\left(\varepsilon_s, \rho_s, d_p, \Theta, e\right) \qquad \mu_s^v = f_4\left(\varepsilon_s, \rho_s, d_p, \Theta^{1/2}, e\right) \tag{7}$$

The solid stress model outlined by Eqs. (6) and (7) will be quoted here as the standard model. Additionally, a general formulation for the solids phase stress tensor that admits a transition between the two regimes is given by :

$$\overline{\overline{S}}_s = \begin{cases} \phi(\varepsilon_f)\overline{\overline{S}}_s^v + \left[1-\phi(\varepsilon_f)\right]\overline{\overline{S}}_s^p & \text{if } \varepsilon_f < \varepsilon^* + \delta \\ \overline{\overline{S}}_s^v & \text{if } \varepsilon_f \geq \varepsilon^* + \delta \end{cases} \tag{8}$$

According to Pannala et al.(2009), two diferent formulations for the weighting parameter "ϕ" can be employed :

$$\phi(\varepsilon) = \frac{1}{1+\upsilon^{\frac{\varepsilon-\varepsilon^*}{2\delta\varepsilon^*}}} \qquad \phi(\varepsilon) = \frac{\text{Tanh}\left(\dfrac{\pi\left(\varepsilon-\varepsilon^*\right)}{\delta\varepsilon^*}\right)+1}{2} \tag{9a,b}$$

In the above equation the void fraction range δ and the shape factor υ are smaller values less than unity. It must be emphasized that when δ goes to zero and ϕ equals to unity, the "switch" model as proposed by Syamlal et al. (1993) based on the Schaeffer (1987) can be recovered. The models based on eqs. (9a) and (9b) will be referred in the numerical simulations as BLEND S and BLEND T, respectivelly.

On the other hand, the Srivastava and Sundaresan (2003), also called "Princeton model", can be placed on the basis of Eq. (9)

Also in equations (4) and (5) $\overline{I}_{fs}$ is the momentum interaction term between the solid and fluid phases, given by

$$\overline{I}_{fs} = -\varepsilon_s\nabla P_f - \beta\left(\vec{v}_s - \vec{v}_f\right) \tag{10}$$

There is a number of correlations for the drag coefficient β (Eqs. 11 to 16). The first of the correlations for the drag coefficient is based on Wen and Yu (1966) work. The Gidaspow drag coefficient is a combination between the Wen Yu correlation and the correlation from Ergun (1952). The Gidaspow blended drag correlation allows controlling the transition from the Wen and Yu, and Ergun based correlations. In this correlation the χ blending function was originally proposed by Lathowers and Bellan (2000) and the value of parameter C controls the degree of transition. From Eq. (14), the correlation proposed by Syamlal and O'Brien (1993) carries the advantage of adjustable parameters C_1 and d_1 for different minimum fluidization conditions. The correlations given in Eq. (15) and Eq. (16) are based on Lattice-Boltzmann simulations. For detailments of these last drag correlations refer to the works by Benyahia et al. (2006) and Wang et al. (2010).

$$\beta_{\text{Wen-Yu}} = \frac{3}{4}C_D\frac{\rho_f\varepsilon_f\varepsilon_s\left|\vec{v}_f - \vec{v}_s\right|}{d_p}\varepsilon_f^{-2.65}$$

$$C_D = \begin{cases} \dfrac{24}{\text{Re}}\left(1+0.15\text{Re}^{0.687}\right) & \text{Re}<1000 \\ 0.44 & \text{Re} \geq 1000 \end{cases} \qquad \text{Re} = \frac{\rho_f\varepsilon_f\left|\vec{v}_f - \vec{v}_s\right|d_p}{\mu_f} \tag{11}$$

$$\beta_{Gidaspow} = \begin{cases} \beta_{Wen\text{-}Yu} & \varepsilon_f > 0.8 \\ \beta_{Ergun} = \dfrac{150\,\varepsilon_s(1-\varepsilon_s)\mu_f}{\varepsilon_f d_p^2} + \dfrac{1.75\,\rho_f\varepsilon_s\left|\vec{v}_f - \vec{v}_s\right|}{d_p} & \varepsilon_f \le 0.8 \end{cases} \tag{12}$$

$$\beta_{Gidaspow\text{-}blended} = \chi\,\beta_{Wen\text{-}Yu} + (1-\chi)\beta_{Ergun} \qquad \chi = \frac{\tan^{-1}\left(C\,(\varepsilon_f - 0.8)\right)}{\pi} + 0.5 \tag{13}$$

$$\beta_{Syamlal\text{-}OBrien} = \frac{3}{4}\frac{\rho_f\varepsilon_f\varepsilon_s}{V_r^2 d_p}\left(0.63 + 4.8\sqrt{\frac{V_r}{Re}}\right)^2 \left|\vec{v}_f - \vec{v}_s\right|$$

$$V_r = 0.5A - 0.03Re + 0.5 \times \sqrt{(0.06Re\,)^2 + 0.12Re\,(2B - A) + A^2} \tag{14}$$

$$A = \varepsilon_f^{4.14} \qquad B = \begin{cases} C_1\varepsilon_f^{1.28} & \varepsilon_f \le 0.85 \\ \varepsilon_f^{d_1} & \varepsilon_f > 0.85 \end{cases}$$

$$\beta_{Hill\text{-}Koch\text{-}Ladd} = 18\mu_f(1-\varepsilon_s)^2\varepsilon_s\frac{F}{d_p^2} \qquad F = f_9(F_0, F_1, F_2, F_3) \tag{15}$$

$$\beta_{Beestra} = 180\frac{\mu_f\varepsilon_s^2}{d_p^2} + 18\frac{\mu_f\varepsilon_f^3\varepsilon_s\left(1 + 1.5\sqrt{\varepsilon_s}\right)}{d_p^2} + 0.31\frac{\mu_f\varepsilon_s\,Re}{\varepsilon_f d_p^2}\frac{\left[\varepsilon_f^{-1} + 3\varepsilon_f\varepsilon_s + 8.4Re^{-0.343}\right]}{[1 + 10^{3\varepsilon_s}\,Re^{-0.5 - 2\varepsilon_s}]} \tag{16}$$

For closing the model, a transport equation for the granular energy Θ provides a way of determine the pressure and viscosity for the solid phase during the viscous regime. Equation (17) is a transport equation for the granular energy Θ. Its solution provides a way of determine the pressure and viscosity for the solid phase during the viscous regime. The terms κ_s γ and ϕ_{gs} are the granular energy conductivity, dissipation and exchange, respectively.

$$\frac{3}{2}\left[\frac{\partial}{\partial t}\varepsilon_s\rho_s\Theta + \nabla\bullet\rho_s\vec{v}_s\Theta\right] = \overline{\overline{S}}_s : \nabla\vec{v}_s - \nabla\bullet(\kappa_s\nabla\Theta) - \gamma + \phi_{gs} \tag{17}$$

$$\kappa_s = f_5\left(\varepsilon_s, \rho_s, d_p, \Theta^{1/2}, e, \beta\right)$$

$$\gamma = f_6\left(\varepsilon_s, \rho_s, d_p, \Theta^{3/2}, e\right) \tag{18}$$

$$\phi_{gs} = f_7\left(\varepsilon_s, \rho_s, d_p, \Theta, \left|\vec{v}_f - \vec{v}_s\right|, \beta\right)$$

In the algebraic approach, instead solving the full equation (17) , the granular energy is obtained by equating the first term on the right hand side with the dissipation term.

The model where Eqs. (5) to (8) and (17) are solved is the kinetic theory model, termed here as KTGF. Conversely, in the constant solids viscosity model (CVM) the solids pressure is defined as in Eq. (6) and the solids viscosity in either plastic and viscous regimes is set equal to a constant.

For erosion calculations in this work we use the monolayer energy dissipation model (Lyczkowski and Bouillard, 2002). In that model the kinetic energy dissipation rate for the solids phase in the vicinity of stationary immersed surfaces is related to erosion rate in m/s by multiplication with an appropriate constant. This constant is function of surface hardness, elasticity of collision and diameter of particles hitting the surface. The kinetic energy dissipation rate Φ_s in W/m^3 for the solids phase is given by :

$$\Phi_s = \left[\varepsilon_s \overline{\overline{\tau}}_s : \nabla \vec{v}_s + \beta \frac{\vec{v}_s^2}{2} \right] \tag{19}$$

3. Numerical method

MFIX (Multiphase Flow with Interphase eXchanges) is an open source CFD code developed at the National Energy Technology Laboratory (NETL) for describing the hydrodynamics, heat transfer and chemical reactions in fluid-solids systems. It has been used for describing bubbling and circulating fluidized beds, spouted beds and gasifiers. MFIX calculations give transient data on the three-dimensional distribution of pressure, velocity, temperature, and species mass fractions.

The hydrodynamic model is solved using the finite volume approach with discretization on a staggered grid. A second order accurate discretization scheme was used and superbee scheme was adopted for discretization of the convective fluxes at cell faces for all equations in this work. With the governing equations discretized, a sequential iterative solver is used to calculate the field variables at each time step. The main numerical algorithm is an extension of SIMPLE. Modifications to this algorithm in MFIX include a partial elimination algorithm to reduce the strong coupling between the two phases due to the interphase transfer terms. Also, MFIX makes use of a solids volume fraction correction step instead of a solids pressure correction step which is thought to assist convergence in loosely packed regions. Finally, an adaptive time step is used to minimize computation time. See Syamlal (1998) for more details. The immersed obstacles were implemented using the cut-cell technique available in the code (Dietiker, 2009)

The numerical runs were based on experiments of Wiman (1994, 1997) in an air pressurized bed with horizontal tubes for four different tube-bank geometries. The T2 and I4 are portrayed in Fig. 1 whereas the S4 and S4D tubes geometry are given in Figure 3. Figure 2 details the domain for the numerical simulations and circumferential angle coordinate for erosion measurement. In Fig. 2a is detailed the mesh for the T2 arrangement This non-uniform stretched grid, follows Cebeci et al. (2005) approach, has fine spacing close to the surface of the tubes and coarse spacing away from the surface. For all the other arrangements the mesh was uniform as depicted in Fig. 2b. The mesh employed for the bed without tube, and the I4 arrangement was 60 × 260 cells, and for the S4 and S4D arrangement 60 × 340. The bed was operated at room temperature (24 C) at pressures between 0.1 and 1.6 MPa and at two different excess velocities: $U_{fl} - U_{mf} = 0.2$ m/s and $U_{fl} - U_{mf} = 0.6$ m/s. Here, U_{fl} is the superficial fluidization velocity based on the free bed cross-section. The voidage at minimum fluidization was 0.46 and the minimum fluidization velocities was 0.42 , 0.31 and 0.18 m/s, for 0.1, 0.4 and 1.6 MPa pressures, correspondingly. The particle diameter and density were 700 μm and 2600 kg/m³. The bubble parameters

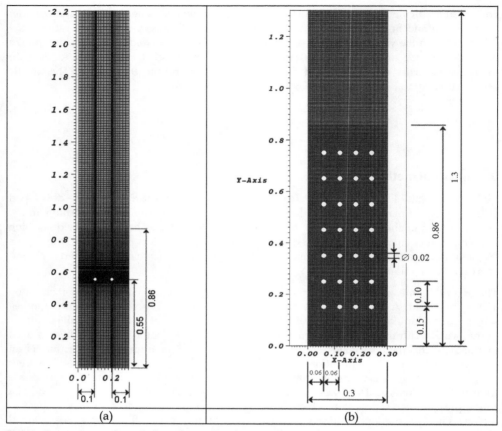

Fig. 1. Geometry and mesh for T2 and I4 tube arrangement.

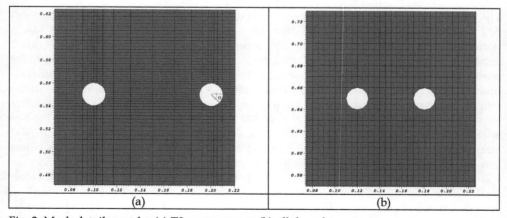

Fig. 2. Mesh detailment for (a) T2 arrangement (b) all the others arrangement

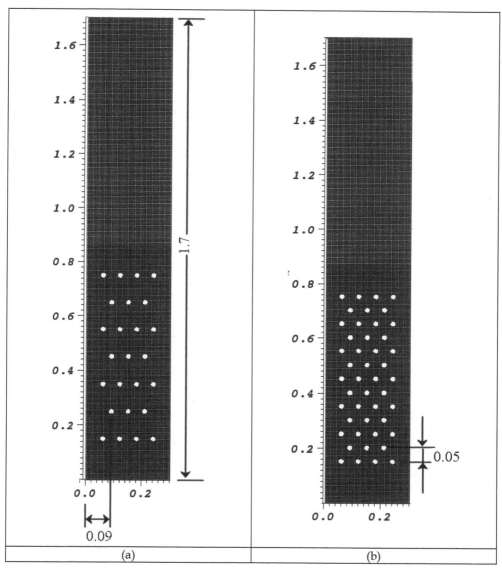

Fig. 3. Geometry and mesh for (a) S4 and (b) S4D tube arrangement

obtained from simulation were based on methodology described in Almstedt (1987), using numerical probes in the domain, centered at (0.15, 0.55) and separated from 15 mm. The target tube for erosion measurements is the one centered at (0.18, 0.55) for all the tube arrangements. More information about the experiments can be accessed from Wiman (1994, 1997).

In this work, the parameters for controlling the numerical solution (e.g., under-relaxation, sweep direction, linear equation solvers, number of iterations, residual tolerances) were kept as their default code values. Moreover, for setting up the mathematical model, when not otherwise specified the code default values were used. The computer used in the numerical simulations was a PC with OpenSuse linux and Intel Quad Core processor. The simulation time was 20 s.

For generating the numerical results, e employed the parameters listed above, referred here as baseline simulation. In addition, for the baseline simulation we employed the Syamlal-O´Brien drag model, the standard solid stress model, and slip and non-slip condition for solid and gas phase, correspondingly. The previous set of models will be referred in the result's section as baseline simulation models.

4. Results and discussion

Figure 4 is a sampling plot showing the instantaneous solids velocities and gas volumetric fraction fields following two different bubbles passage around the obstacle. Analysis of Fig. (4) shows the bubble splitting mechanism taking place and the characteristic time scale for the bubble passage. After the bubble passage, the solid wake has higher solid velocity magnitude around the obstacle. The solids upward movement following the bubble wake and wall downward movement is kept unchanged. Generally, for beds without internal obstacles, bubble size increases with bed height, particle size and superficial velocity. Analysis of CFD results shows that the presence of tubes was found to alter such general trends for bubble growth.

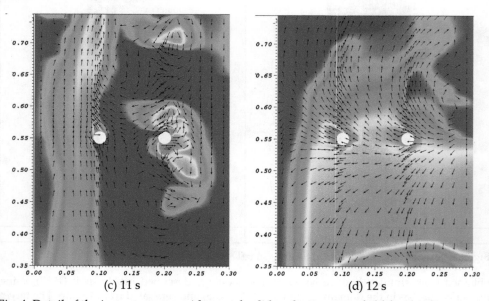

(c) 11 s   (d) 12 s

Fig. 4. Detail of the instantaneous voidage and solids velocity vector field for the T2 arrangement

Figure 5 presents the time averaged kinetic energy dissipation as a function of circumferential position θ on the tube surface. As it would be seen from the majority of

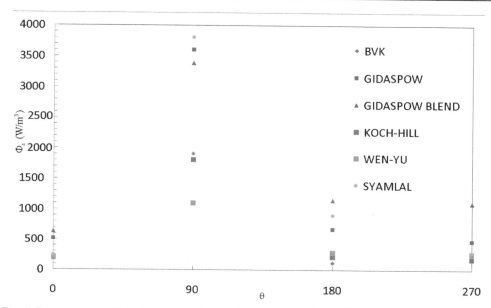

Fig. 5. Time averaged kinetic energy dissipation predicted for different gas-solid drag models – T2 arrangement

results the most severe dissipation occurs on the lower parts of the tube for an angle corresponding to 90 degrees. This fact is in agreement with experimental measured values of erosion from Wiman (1994). Analysis of results shows that, except for the BVK drag model, all the drag models predict the highest dissipation rate occurring at 90 degrees. The value predicted using the Syamlal-O'Brien drag model is the highest. There is a noticeable difference in the value of dissipation value between the Gidaspow and Gidaspow blend drag models, whereas the results by the Wen-Yu drag locate in the intermediate range. By his turn, the results by HYS and Koch-Hill models locates near the Wen-Yu values.

Also, analysis of transient simulated results shows the highest values of kinetic energy dissipation rate occur when the particle fraction suddenly changes from a low to a high value, which corresponds to the tube being hit by the wake of a bubble.

Figure 6 presents the result of kinetic energy dissipation for different solid stress models. As shown the peak values of the dissipation rate are close to 180 degrees for the model with (baseline) and without the blending function discussed in section 2. On the other hand, the maximum value predicted by the constant solids viscosity model (NO KTGF) locates around 180 degrees, and it is superior to the predicted by the solids kinetic energy theory.

Figure 7 presents the result of kinetic energy dissipation for different tube surface slip conditions. The baseline case considers the free slip condition for the solids and non-slip condition for the gas. The peak values do not change when considering the solids with the non-slip condition. On the other hand, when considering slip conditions for both phases the value decreased. The above results, suggests the free slip condition for the gas phase

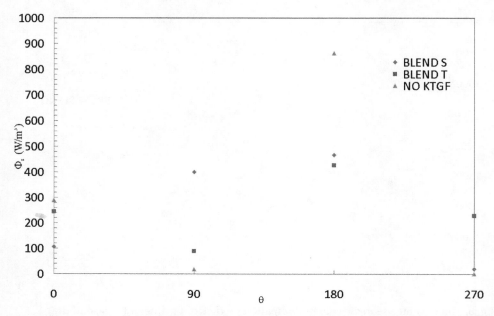

Fig. 6. Time averaged kinetic energy dissipation predicted for different solid stress models

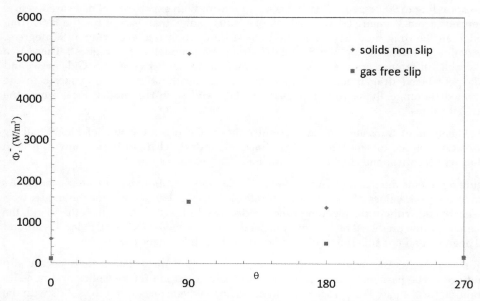

Fig. 7. Time averaged kinetic energy dissipation predicted for different tube surface slip conditions

on the surface of the tube plays an important role on the value of the kinetic energy dissipation.

Figures 8 and 9 shows a comparison between the numerically predicted time averaged values of the kinetic energy dissipation rate, using the baseline simulation models discussed in section 2, and those based on the experimental values for the T2 arrangement in the work of Gustavson and Almstedt (2000). A comparison of the numerical and experimental results for the two different operational pressures (c.f. Figs 8 and 9) shows some degree of discordance. However, the lack of minutely agreement with experimental values, the results can be compared for recognizing similar drifts. For instance, from the Fig 8, the higher simulated values of the dissipation rate occurring in the bottom position of the tube, i.e. for $\theta < 180°$, is in agreement with the experimental counterpart. Similarly, the increase of dissipation rate with increased operational pressure for the simulated results is in agreement with the experimental values. Regarding the erosion and baseline models used for the simulation, some remarks towards better agreement with experimental values can also be done. According to the monolayer erosion model and its discussion above Eq. (19) some degree of uncertainty is associated to the multiplying constant, as the exact value of elasticity of collision is not known. It is also expected, that adjustments in the baseline simulation models, such as those outlined in Figs. (5) to (7) would produce better agreement.

Figure 10 is a sampling plot showing the instantaneous gas volumetric fraction fields for different tube arrangements. Analysis of Fig. (10) shows the influence of immersed obstacles on the bubble splitting mechanism taking place and the bubble passage pattern. Above the tube bank, the bubble appears to grow to size similar to the without tube geometry. For the geometry with tubes the bubble encompasses the obstacles but not at the full width of the bed. The interaction is stronger for the denser tube geometry.

Figures 11 and 12 shows a comparison between the numerically predicted values of bubble frequency, using the baseline simulation models discussed in section 2, and those based on the experimental measurements from the works of Almstedt (1987) and Wiman (1995). As shown in the Fig. 10, the calculated values of N_b are underestimated at higher pressures, while at low pressures, there is a quite good agreement between calculated and experimental results. This conclusion holds true both for the I4 and for the S4D tube arrangement. As in the experimental results there are no noticeable differences for the two tubes arrangements.

Figures 13 and 14 shows a comparison for the bubble frequency for the bed without tube and with the S4 arrangement. As shown in Fig. 13 for the bed without tubes the trend points to a frequency agreement between 0.4 and 0.6 MPa. Up this range the values differences increases as pressures increases up to 1.6 MPa. For the S4 arrangement the trend is similar to the I4 and S4D arrangement. The experimental results depicted in Figs. 11 to 14 suggest that the frequency increases with pressure both with and without tubes. For numerical results this holds true only for low pressures, i.e., 0.1 and 0.4 MPa. The numerical results suggest a maxima occurring between 0.4 and 1.6 MPa. On the other hand, the numerical results corroborate the experimental trend that the mean frequency is higher for the bed with tubes than for the freely bubbling bed.

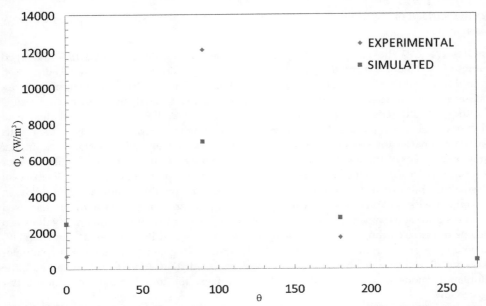

Fig. 8. Simulated kinetic energy dissipation at various circumferential angular positions in the surface of the tube and values from experimental results by Gustavsson and Almstedt, 2000. Operational pressures : 0.8 MPa ;

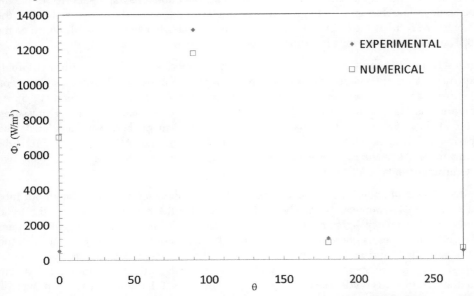

Fig. 9. Simulated kinetic energy dissipation at various circumferential angular positions in the surface of the tube and values from experimental results by Gustavsson and Almstedt, 2000. Operational pressures : 1.6 MPa

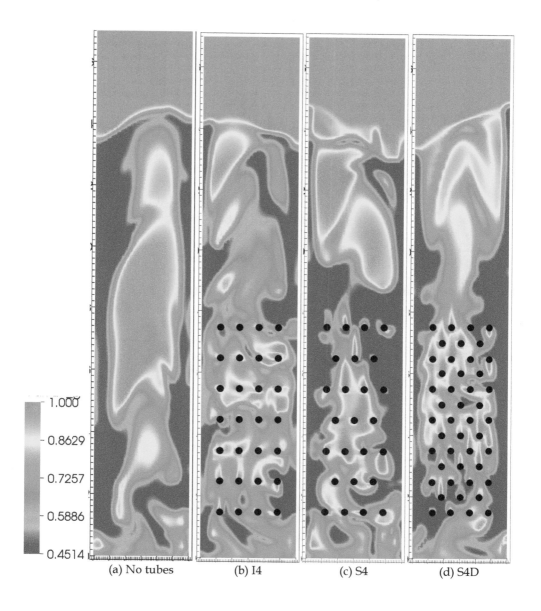

1.000

0.8629

0.7257

0.5886

0.4514

(a) No tubes   (b) I4   (c) S4   (d) S4D

Fig. 10. Snapshots of voidage field at 4 s for different tube arrangements. P = 0.4 MPa , Uf = 0.6 m/s

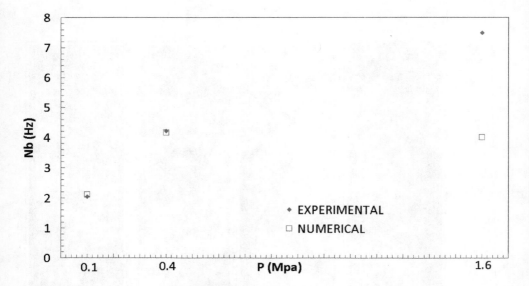

Fig. 11. Nb versus pressure, numerical X experimental: I4 arrangement, $U_f = 0.6$ m/s

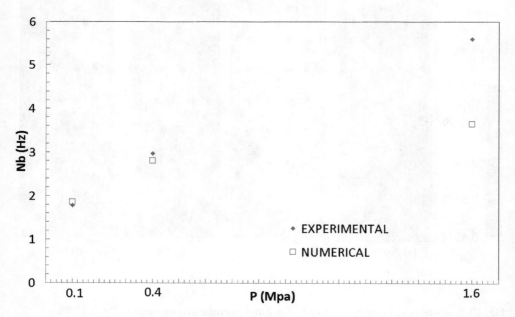

Fig. 12. Nb versus pressure, numerical X experimental: S4D arrangement, $U_f = 0.6$ m/s

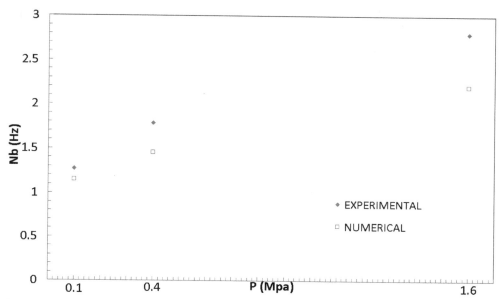

Fig. 13. Nb versus pressure, numerical X experimental: no tubes, $U_f = 0.6$ m/s

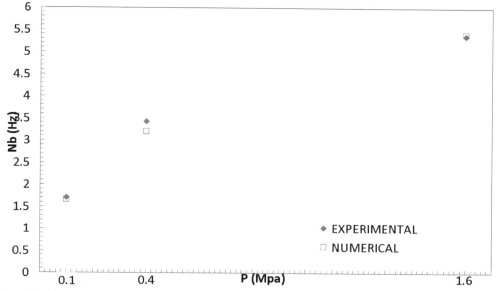

Fig. 14. Nb versus pressure, numerical X experimental: S4 arrangement, $U_f = 0.6$ m/s

Figure 15 to 18 shows a comparison for the mean bubble velocity. In all cases, the simulated results are underestimated in relation to the experiments. However, the trend observed for the experimental results with a maxima around 0.4 MPa is verified for the numerical results for all the tube arrangements. For the bed without tubes the experimental increase trend of Vb with pressure is valid for pressures higher than 0.4 MPa.

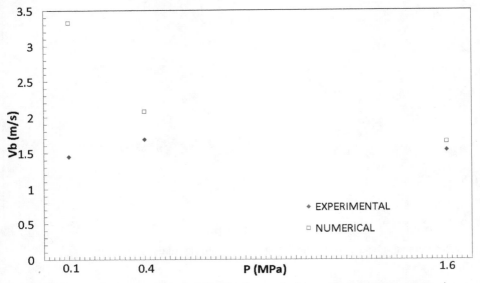

Fig. 15. Vb versus pressure, numerical X experimental: : I4 arrangement, U_f = 0.6 m/s

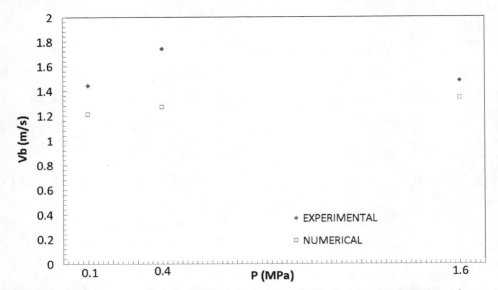

Fig. 16. Vb versus pressure, numerical X experimental: S4D arrangement, U_f = 0.6 m/s

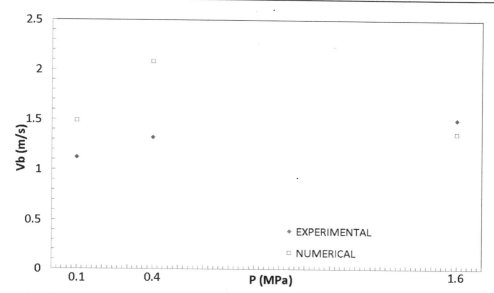

Fig. 17. Vb versus pressure, numerical X experimental: : no tubes, U_f = 0.6 m/s

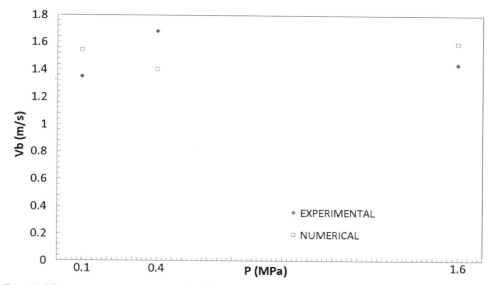

Fig. 18. Vb versus pressure, numerical X experimental: : S4 arrangement, U_f = 0.6 m/s

Figures 19 and 20 shows a comparison for the bubble frequency for U_f = 0.2 m/s. Comparison with results in Figs. 12 and 14 shows that the numerical results, although smaller are closer to the experimental. The tendency holds true for both S4 and S4D arrangements. Comparison with the numerical results for the S4D and S4 arrangements

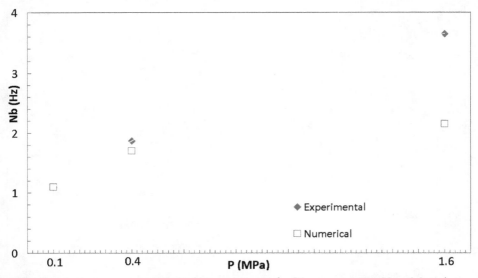

Fig. 19. Nb versus pressure, numerical X experimental: : S4 arrangement, U_f = 0.2 m/s

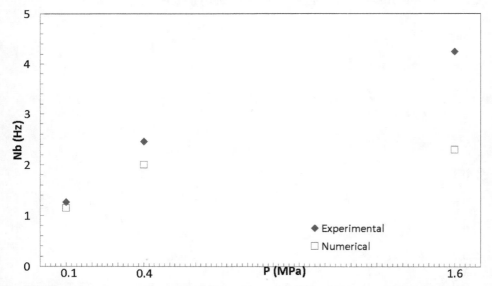

Fig. 20. Nb versus pressure, numerical X experimental: S4D arrangement, U_f = 0.2 m/s

given in Fig. 6 and 7, also shows that Nb increases with increasing excess velocity. The last, is the same drift verified for the experimental values.

Figures 21 and 22 present the results for the time averaged kinetic energy dissipation as a function of circumferential position θ on the tube surface for the S4 arrangement for two distinct pressures. As it would be seen the highest experimental results are between 90 and

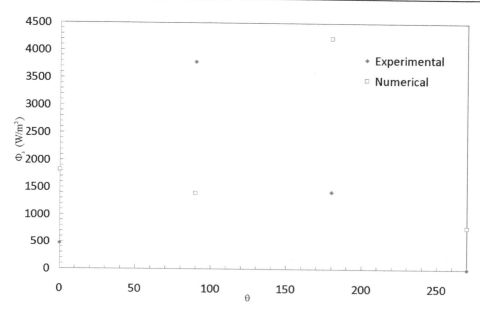

Fig. 21. Time averaged kinetic energy dissipation predicted for different tube arrangements at two different operating pressures: : S4 arrangement, P = 0.1 MPa

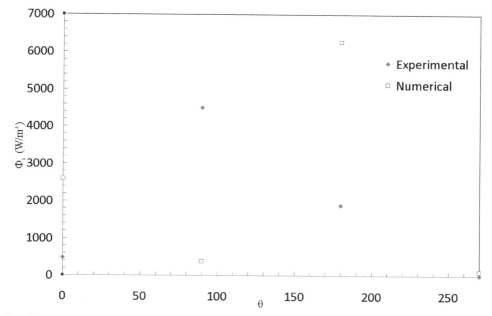

Fig. 22. Time averaged kinetic energy dissipation predicted for different tube arrangements at two different operating pressures : : S4 arrangement, P = 1_6 MPa

240 degrees, while for the numerical are between 30 and 150. The numerical values are over predicted in the range 0 to 100 degrees and above 240 degrees. In the range from 100 to 240 degrees the numerical are below the experimental. Similar trends are verified for the 1.6 MPa pressure, with the experimental curves less sensitive to pressure variation . By his turn the numerical values, show more sensitivity to pressure, although with the same magnitude order.

5. Conclusion

In this work was investigated numerically the hydrodynamics of two dimensional beds with immersed tubes. The simulations were based on an experimental bed with different tube bank geometries, operating pressures and gas excess velocities. The objective of this study was two fold: explore and investigate some effects not previously explored in the literature, to verify the feasibility of the MFIX code for such a kind of study.

The simulation's results were framed in terms of averaged solids kinetic energy dissipation rate and bubble parameters (frequency and mean velocity). A comparison between the numerical results for frequency and the experiments shows good agreement for low pressures. Also, some similar drifts were identified, e.g., greater frequency for greater excess velocities. By his turn, the bubble velocity numerical results agrees better with experiments for high pressures. For the bed with tubes, a maxima of bubble velocity for intermediate pressure is identifiable, this could not be verified for the numerical results. Our results points to significant influences in the predicted dissipation rates and consequently, in the erosion rate, when employing different drag models. The dissipation rate is also influenced by either the use of blending functions for the transition between the plastic and viscous regime of solids flow or the use of a constant viscosity model. The results are key sensitive to the slip condition for the gas phase in the surface of the tube. In the case of a free slip condition for the gas phase the lowest values of dissipation are obtained By his turn, for the solids kinetic dissipation energy, the range of angles that gives maximum values are shifted in relation to the experiments. Finally, remarks towards better agreement with experimental values can also be done for the energy dissipation model. Specifically, according to the monolayer erosion model and its discussion above Eq. (19) some degree of uncertainty is associated to the multiplying constant, as the exact value of elasticity of collision is not known.

6. References

Achim, D., Easton, A. K., Schwarz, M. P., Witt, P.J., Zakhari, A., 2002, "Tube erosion modelling in a fluidised bed", Applied Mathematical Modelling, Vol 26, pp. 191-201.

Almstedt, A. E., 1987, A study of bubble behaviour and gas distribution in pressurized fluidized beds burning coal. Thesis for the degree of Licentiate of Engineering, Chalmers University of Technology, Goteborg. Sweden.

Anderson, T. B., 1967, "A fluid mechanical description of fluidized beds: Equations of motion", Industrial Engineering Chemical Fundamentals, Vol 6, pp. 527-539.

Benyahia, S., 2008, "Validation study of two continuum granular frictional flow theories", Industrial Engineering Chemical Research, 47, 8926-8932.

Benyahia, S., Syamlal, M., O'Brien, T. J., "Summary of MFIX Equations 2005-4", 1 March 2006, Available from http://www.mfix.org/documentation/MfixEquations2005-4-1.pdf.

Benyahia, S., Syamlal, M., O'Brien, T. J., 2006, "Extension of Hill–Koch–Ladd drag correlation over all ranges of Reynolds number and solids volume fraction", Powder Technology, 162, 166-174.

Bouillard, J. X., Lyczkowski, R. W., 1991, "On the erosion of heat exchanger tube banks in fluidized-bed combustor", Powder Technology, Vol 68, pp. 37-51.

Cebeci, T., Shao, J. P., Karyeke, F., Laurendeau, E., 2005, Computational Fluid Dynamics for Engineers, Horizon Publishing Inc, USA, 402p.

Dietiker, J., "Cartesian Grid User Guide", 4 September 2009, Available from https://mfix.netl.doe.gov/documentation/Cartesian_grid_user_guide.pdf.

Enwald, H., Peirano, E., Almstedt, A. E., Leckner, B., 1999, "Simulation of the fluid dynamics of a bubbling fluidized bed: Experimental validation of the two-fluid model and evaluation of a parallel multiblock solver", Chemical Engineering Science, 54, 311-328.

Ergun, S., 1952, "Fluid-flow through packed columns", Chemical Engineering Progress, Vol 48, n. 2, pp. 91-94.

Fan, J. R., Sun, P., Chen, L. H., Cen, K. F., 1998, "Numerical investigation of a new protection method of the tube erosion by particle impingement", Wear, Vol 223, pp. 50-57.

Gustavsson, M., Almstedt, A. E., 1999, "Numerical simulation of fluid dynamics in fluidized beds with horizontal heat exchanger tubes", Chemical Engineering Science, 55, 857-866.

Gustavsson, M., Almstedt, A. E., 2000, "Two-fluid modelling of cooling-tube erosion in a fluidized bed", Chemical Engineering Science, 55, 867-879.

He, Y. R., Lu, H. L., Sun, Q. Q., Yang, L. D., Zhao, Y. H., Gidaspow, D., Bouillard, J., 2004, "Hydrodynamics of gas-solid flow around immersed tubes in bubbling fluidized beds", Powder Technology, 145, pp. 88-105.

He, Y. R., Zhan, W., Zhao, Y., Lu, H. Schlaberg, I., 2009, "Prediction on immersed tubes erosion using two-fluid model in a bubbling fluidized bed", Chemical Engineering Science, 64, pp. 3072-3082.

Kobayashi, N., Yamazaki, R., Mori, S., 2000, "A study on the behaviour of bubbles and solids in bubbling fluidized beds", Powder Technology, Vol 113, pp. 327-344.

Lathowers, D., Bellan, J., 2000, "Modeling of dense gas-solid reactive mixtures applied to biomass pyrolysis in a fluidized bed", Proceedings of the 2000 U.S. DOE Hydrogen Program Review, NREL/CP-570-28890. USA.

Lee, S. W., Wang, B. Q., 1995, "Effect of particle-tube collision frequency on material wastage on in-bed tubes in the bubbling fluidized bed combustor", Wear, Vol 184, pp. 223-229.

Lyczkowski, R. W., Bouillard, J. X., 2002, "State-of-the-art review of erosion modeling in fluid/solid systems", Progress in Energy and Combustion Science, Vol 28, pp. 543-602.

Ozawa, M., Umekawa, H., Furui, S., Hayashi, K., Takenaka, N., 2002, "Bubble behavior and void fraction fluctuation in vertical tube banks immersed in a gas-solid fluidized-bed model", Experimental Thermal and Fluid Science, Vol 26, pp. 643-652.

Pannala, S., Daw, C. S., Finney, C. E. A., Benyahia, S., Syamlal, M., O'Brien, T. J., "Modelling the collisional-plastic stress transition for bin discharge of granular material", 2009, Powders and Grains 2009 – Proceeding of the 6th International Conference on Micromechanics of Granular Media, pp. 657-660.

Schaeffer, D. G., 1987, "Instability in the evolution equations describing incompressible granular flow", Journal Differential Equations, v. 66, pp. 19-50.

Siravastava, A., Sundaresan, S., 2003, "Analysis of a frictional–kinetic model for gas–particle flow", Powder Technology, v. 129, pp. 72-85.

Syamlal, M., 1998, "MFIX Documentation, Numerical Techniques", Technical Note, DOE/MC-31346-5824, NTIS/DE98002029, National Technical Information Service, Springfield, VA, USA.

Syamlal, M., Rogers, W. A., O'Brien, T. J., 1993, "MFIX Documentation, Theory Guide", Technical Note, DOE/METC-94/1004, NTIS/DE94000087, National Technical Information Service, Springfield, VA, USA.

Wang, J., van der Hoef, M. A., Kuipers, J. A. M., 2010,"CFD study of the minimum bubbling velocity of Geldart A particles in gas-fluidized beds", Chemical Engineering Science, 65, pp. 3772-3785.

Wen, C. Y., Yu, Y. H., 1966, "Mechanics of Fluidization", Chemical Engineering Progress Symposium Series, Vol 62, n. 62, pp. 100-111.

Wiman, J., 1994, An experimental study of hydrodynamics and tube erosion in a pressurized fluidized with horizontal tubes. Thesis for the degree of Licentiate of Engineering, Chalmers University of Technology, Goteborg. Sweden.

Wiman, J., Almstedt, A. E., 1997, "Hydrodynamics, erosion and heat transfer in a pressurized fluidized bed: influence of pressure, fluidization velocity, particle size and tube bank geometry", Chemical Engineering Science, Vol 52, pp. 2677-2695.

Wong, Y. S., Seville, J. P. K., 2006, "Single-particle motion and heat transfer in fluidized beds", AIChe Journal, Vol 52, pp. 4099-4109.

Systematic Framework for Multiobjective Optimization in Chemical Process Plant Design

Ramzan Naveed[1], Zeeshan Nawaz[2],
Werner Witt[3] and Shahid Naveed[1]
[1]Department of Chemical Engineering,
University of Engineering and Technology, Lahore
[2]Chemical Technology Development, STCR,
Saudi Basic Industries Corporation (SABIC)
[3]Lehrstuhl Anlagen und Sicherheitstechnik,
Brandenburgicshe Technische Universität, Cottbus
[1]Pakistan
[2]Kingdom of Saudi Arabia
[3]Germany

1. Introduction

For solving multiobjective decision making problems, a systematic and effective procedure is required. As far as the process or control system has to be modified process simulators like Aspen Plus™, Aspen Dynamics are widely used. But these simulators are not designed for investigation of other objectives as environment and safety. Due to complex and conflicting nature of multiobjective decision making an integrated optimization tool should be of value. In this chapter a systematic methodology based on independent modules and its different stages to deal this problem is presented in detail.

2. Proposed methodology

The methodology is built around several standard independent techniques. These techniques have been suitably modified/adopted and woven together in an integrated plate form. The main aim is to standardize the screening and selection of decisions during design/modification of chemical process plant and optimizing the process variables in order to generate a process with improved economics along with satisfaction of environmental and safety constraints. The methodology (see Figure 1) consists of four layers/stages:

- Generation of alternatives and problem definition;
- Analysis of alternatives i.e. generation of relevant data for comparison of Environmental, economic and safety objectives
- Multiobjective decision analysis/ optimization
- Design evaluation stage i.e. decision making from the pareto-surface of non-inferior solution or ranking of alternatives

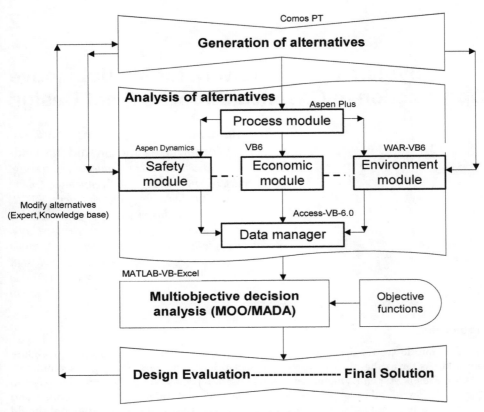

Fig. 1. Simplified block diagram of proposed methodology

2.1 Stage I: Generation of alternatives and problem definition

The first layer composed of following tasks:

Definition of the scope of the study,
Statement of key assumptions and the performance targets such as quality etc.,
Degree of freedom analysis,
Identification of the key design, control, and manipulated variables,
Definition of the system boundary,
Identification of constraints,
Choice of functional unit for all calculations,
Collection of relevant information about process and chemicals to be handled,
Generation of different alternatives either based on suggestion from independent departments or using the individual objective modules from stage II.

Seader et al. (1999)[6] has described rules for selection of process variables in the book "Process design principles-synthesis, analysis and evaluation". The data and information about the process and chemicals involved such as thermodynamic and kinetic data can be found from journal articles, patents or handbooks. Current chemical prices can be obtained from market reports if not available in main plant documentation or company central data base. In addition to these sources, some data related to quantification of environmental impacts and material safety data sheets of chemical are also collected from commercial data bases so that an impact assessment and safety analysis can be performed in subsequent design steps. Commercial computer aided tool like Comos[PT] can be used for plant documentation and to support stage-I of proposed methodology.

2.2 Stage II: Analysis of alternatives

This stage is composed of independent modules used to generate relevant information for evaluation of economic, safety and environmental performance objectives. These modules are:

- Process module
- Safety module [1,3]
- Economic module
- Environment module [2]
- and a data manager for managing the relevant information generated from these modules.

2.3 Process module

In the process module, an operation model of the process system has to be developed for evaluating alternatives. The configured simulation model has to be able to reproduce the selected results to an accepted degree of accuracy. This simulation model can be used for design and operation, revamping and debottlenecking of the process under study[7]. Three major integrated simulation systems widely used in the firms and companies for this purpose are Aspen technology (Aspen Plus, Aspen dynamics etc), Hyprotech (Hysys process, Hysys plant etc) and Simulation Sciences (Pro/II etc.). Aspen Plus[TM] 12.1 is used in this work for development of simulation model and linked in a visual basic platform for integration with safety, economic and environment modules. The most important results available from the process simulation model are material and energy balance information for both streams and units, rating performance of units and tables and graphs of physical properties. A brief description of Aspen Plus[TM] 12.1 and steps involved in development of the process simulation model is described here below.

Aspen Plus[TM]

Aspen Plus[TM] supports both sequential modular and equation oriented computation strategy and allows the user to build and run a steady-state simulation model for a chemical process. It provides a flexible and productive engineering environment designed to maximize the results of engineering efforts, such as user interface mode manager, quick property analysis, rigorous and robust flowsheet modelling, interactive architecture, powerful model analysis tools and analysis and communication of results. Therefore, it lets the user to focus his/her energies on solving the engineering problems, not on how to use the software. It is not only good for process simulation but also allows to perform a wide range of other tasks such as estimating and regressing physical properties, generating custom graphical and tabular output results, sensitivity analysis, data-fitting plant data to simulation models, costing the plant, optimizing the process, and interfacing results to spreadsheets.

The development of a simulation model for a chemical process using Aspen Plus™ 12.1 involves the following steps (see details in table 1):

1. Define the process flowsheet configuration by specifying
    • Unit operations
    • Process streams flowing between the units
    • Unit operation models to describe each unit operation
2. Specify the chemical components,
3. Choose a thermodynamic model to represent the physical properties of the components and mixtures in the process,
4. Specify the component flow rates and thermodynamic conditions (i.e. temperature, pressure, or phase condition) of the feed streams,
5. Specify the operating conditions for the unit operations,

Step	Used to
Defining the flowsheet	Break down the desired process into its parts: feed streams, unit operations, and product streams
Specifying stream properties and units	Calculate the temperature, pressure, vapor fraction, molecular weight, enthalpy, entropy and density for the simulation streams
Entering components	From a databank that is full of common components
Estimating property parameters	Property Constant Estimate System (PCES) can estimate many of the property parameters required by physical property models
Specifying streams	Streams connect unit operation blocks in a flowsheet and carry material and energy flows from one block to another. For all process feed streams, we must specify flowrate, composition, and thermodynamic condition
Unit operation blocks	We choose unit operation models for flowsheet blocks when we define our simulation flowsheet

Table 1. Developmental process for an Aspen Plus™ simulation model

2.4 Safety module

Safety module is based on combination of conventional standard risk analysis techniques and process disturbance simulation. This module not only generates relevant information related to safety aspects for multiobjective decision analysis but also used for safety/risk analysis and optimization. The purpose of this module is to determine risk from operational disturbances and to develop effective risk reductions. It can be divided into the following steps (Figure 2):

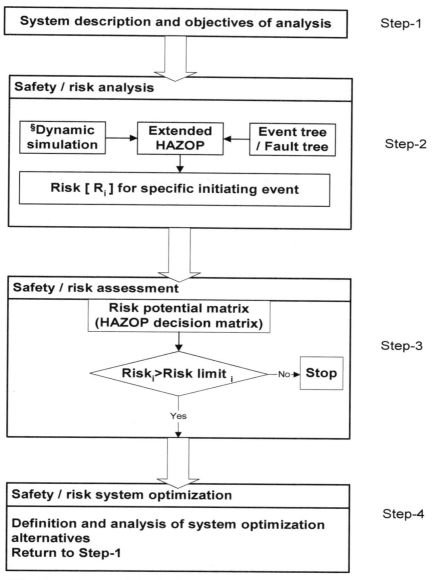

§ Simulation of process related malfunctions

i = { financial risk , environmental risk , human health risk }

Step 1: System description and objectives of analysis (before starting safety and risk analysis)

Step 2: Safety/risk analysis (identification of weak points via Extended HAZOP)

Step 3: Safety/risk assessment (categorization of risk via risk potential matrix (HAZOP decision matrix))

Step 4: Safety/risk system optimization

Fig. 2. Simplified block diagram of safety module

2.4.1 Step 1: (Before starting safety/risk analysis) - Description of system and objectives of analysis

For efficient safety/risk studies, the analyst must have an accurate description of the system to be investigated and a clear objective of the analysis study. Therefore, in this step the purpose, objectives, and scope of the study are clearly defined. The necessary information required for the study such as process flow diagrams, piping and instrumentation diagrams, plant layout schematics, material safety data sheets, equipment data sheets, operating instructions, start up and emergency shutdown procedures, and process limits, etc. is gathered from plant documentation. A team under a trained and experience leader with five to seven people including experts of the design and operation of the subject process may be formulated.

2.4.2 Step 2: Safety/risk analysis (Identification of weak points via extended HAZOP) - Extended HAZOP

Our intention is to identify weak points due to disturbances in operation, which may or may not be hazardous, in order to improve safety, operability, and/or profit at the same time. Extended HAZOP (HAZOP supported by dynamic simulation, event tree and fault tree techniques and HAZOP decision matrix) is used not only for identification of weak points but also for generation and analysis of optimization proposals [8-11]. Extended HAZOP differs from the standard HAZOP approach in following aspects:

i.   Use of dynamic simulation:

In Extended HAZOP, the analysis of the influence of disturbances (failures) on the behaviour of the process is based on shortcut or simplified hand calculations or dynamic simulation. Aspen dynamics is used for this purpose.

ii.   Classification of risk related consequences:

Each established consequence (hazard) has to be expressed by a consequence class (C). The plant specific scoring (from 0 (lowest) to 8 (highest)) chart is given in Table 2 (a & b) based on principle consequence analysis. For classification of consequences based on principle release estimates, accident consequence analysis techniques (models for calculation of toxic, fire and explosion effects) and plant location data (capital investment, population density etc.) have to be considered.

Illustrative Example 1

Figure 3 shows the plant lay out considered for developing plant specific consequence scoring chart. The area around the plant is open fields (rural condition). As weather conditions changed around the year, so certain assumptions are made to results in worse case conditions for consequence analysis. These include weather conditions and wind speed that result in smallest value of dispersion coefficients. Therefore, stability "F" and wind speed as low as possible (1.5 m/s) is selected. It is assumed that 10 workers are present (working 24 h each day), which are not distributed uniformly, on the land in area (100 m x 100 m) around the column under study. Acetone is selected as representative fluid for consequence analysis.

Acetone vapours released from the vent line at a rate of 1616 kg/h due to loss of cooling medium. It is assumed that released vapours form a cloud for 30 minutes before being

Effects	Class	Financial loss (€)	Class related consequences: examples
Function impairment	*	< 10	: Product quality lowering (brief)
	1	$10^1 - 10^2$	: Product quality lowering
	2	$10^2 - 10^3$	: Product quality lowering (long term)
Functional Loss	3	$10^3 - 10^4$	: Production disturbance (brief) Soil contamination Safe dispersion of material release from vent line
	4	$10^4 - 10^5$	: Production disturbance Material release from the piping Pump damage (pressure impacts)
	5	$10^5 - 10^6$	: Production disturbance (long term) Jet fire as result of release of material from vent line Pool fire (from pump leakage)
Safety and Environmental pollution	6	$10^6 - 10^7$	Fireballs due to catastrophic rupture of vapour product line
	7	$10^7 - 10^8$	Vapour cloud explosion (ignoring domino effect)
	8	$>10^8$	Vapour cloud explosion along with domino effect

Effects	Class	Community	Class related consequences: examples
Function impairment	*	No effect on people	: Product quality lowering (brief)
	1	Nuisance effect	: Product quality lowering
	2	Minor irritation effect to people & local news	: Product quality lowering (long term)
Functional Loss	3	Moderate irritation effect to people and non compliance to laws, local news	: Production disturbance (brief) Soil contamination Safe dispersion of material release from vent line
	4	Moderate irritation effects to people & environment, single injuries and regional news	: Production disturbance Material release from the piping Pump damage (pressure impacts)
	5	Significant effects to people and environment, > 1 injuries & regional news	: Production disturbance (long term) Jet fire as result of release of material from vent line Pool fire (from pump leakage)
Safety and Environmental pollution	6	Major effects to people and environement, multiple injuries, fatality likely, regional news	Fireballs due to catastrophic rupture of pipe or condenser (vapour product line)
	7	Severe effects to people and environment, fatality, regional news	Vapour cloud explosion (ignoring domino effect)
	8	Multiple fatalities and process shutdown certain, international news	Vapour cloud explosion along with domino effect

Environment and Health consequences

Table 2. Scoring chart for Consequence Financial consequences [3.4]

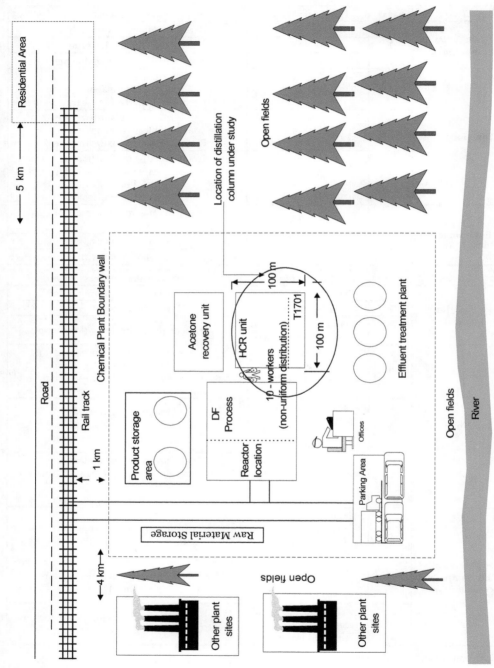

Fig. 3. Plant lay out for establishing consequence score chart

ignited and leads to vapour cloud explosion. The physical effects of this scenario or event is calculated as:

Weight of fuel in the cloud= M = 1616 / 2 = 808 kg

Then amount of TNT equivalent to the amount of this flammable material is

$$M_{TNT} = \alpha \cdot \frac{M \cdot H_c}{H_{TNT}}$$

Where α = explosion efficiency ~ 0.05 (Cameron 2005) H_c = heat of combustion of fuel ~ 3.03 x 104 kJ/kg for acetone

H_{TNT} = TNT blast energy ~ 5420 kJ/kg; so M_{TNT} = 225.25 kg

Then, using relation $Z = R / (M_{TNT})^{1/3}$ and figure 4, scaled distance and overpressure is estimated. Table 3 presents the results obtained.

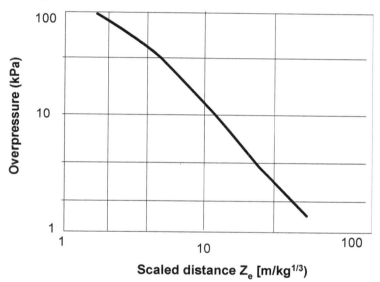

Scaled distance Z$_e$ [m/kg$^{1/3}$)

Fig. 4. Overpressure versus scaled distance for TNT explosions on flat surfaces (Tweeddale 2003, p. 115)

Distance , R M	Scaled distance, Z m / kg1/3	Overpressure, Δp kPa
10	1.64	90
20	3.28	40
50	8.21	20
100	16.40	7

Table 3. Results of physical effects of vapour cloud explosion

It is estimated that severe structural damage and 15 % chance of fatality outdoors or 50 % chance indoor will be experienced out to 20 m and almost complete destruction of all ordinary structures and 100 % chance of fatality indoors to 10 m distance.[8] (see Cameron 2005, p. 268).

iii.   Classification of frequencies of risk related consequences:

The frequency of occurring for each possible consequence (hazard) has to be expressed by a frequency class, called (F) according to the scoring chart for frequency (Table 3.4): Definition of frequency class may be supported by Event Tree and/or Fault tree analysis techniques or Layer of protection analysis (LOPA) or historical databases.

For establishing frequency class: Estimation / calculation of frequency of vapour cloud explosion and fatality of person because of release of material due to catastrophic rupture of distillation column.

Frequency of catastrophic rupture of column = 10-6 (Taken from table 4)

Class	Frequency of occurring incident		
	Frequency 1/y	Comprehension	Examples based on general data bases
9	$<10^{-8}$	Very very small	Catastrophic rupture or leakage of pipe of diameter > 150 mm
8	10^{-8} - 10^{-7}	Very small	Catastrophic rupture of pipe of diameter ≤ 50 mm
7	10^{-7} - 10^{-6}	Small	Catastrophic rupture of fractionating system (excluding piping), storage tank rupture
6	10^{-6} - 10^{-5}	Less small	Pipe residual failure, 100 m full breach, Double wall tank leakage
5	10^{-5} - 10^{-4}	Moderate	Process vessel leakage of ≥ 1 mm diameter
4	10^{-4} - 10^{-3}	Less moderate	Pump leakage , Heat exchanger leakage
3	10^{-3} - 10^{-2}	Less high	Safety valve open spuriously, Large external fire
2	10^{-2} - 10^{-1}	High	Cooling water failure, BPCS instrument loop failure
1	10^{-1} - 10^{0}	Very high	Operator failure, Regulator failure , Solenoid valve failure
*	$>10^{0}$	Very very high	Power failure in developing countries, Operators failure under high stress

Table 4. Scoring chart for frequency [3.4]

Probability of ignition of released material = 0.10 (CCPs 2000, Borysiewich 2004)

Probability of VCE if released material ignited = 0.01 (CCPs 2000, Borysiewich 2004)

Probability of fatality of a person exposed to overpressure of 40 kPa due to VCE = 0.20 (Tweeddale 2003, p. 117 Figure 5-14)

Then, frequency of vapour cloud explosion $= 10^{-6} \cdot 0.10 \cdot 0.01 = 10^{-9}$

So frequency class for this scenario $= 9$

Frequency of fatality of a person exposed to VCE $= 10^{-9} \cdot 0.20 = 2.10^{-10}$

So frequency class for this scenario $= 9$

iv. Way of documenting the HAZOP results

The Extended HAZOP methodology worksheet for documenting the HAZOP team results is shown in Figure 5. Below consequence the physical effects and risk has to be documented first and next risk has to be classified using score charts (Table 3.2) related to financial, environment and health related consequences. The worst score of each risk has to be documented. For each risk related consequence, frequency class has also to be established.

Consequence, €		<10	$10^1 - 10^2$	$10^2 - 10^3$	$10^3 - 10^4$	$10^4 - 10^5$	$10^5 - 10^6$	$10^6 - 10^7$	$10^7 - 10^8$	$>10^8$
Frequency $1/y$	C F	*	1	2	3	4	5	6	7	8
$>10^0$	*									
$10^{-1} - 10^0$	1									
$10^{-2} - 10^{-1}$	2									
$10^{-3} - 10^{-2}$	3									
$10^{-4} - 10^{-3}$	4									
$10^{-5} - 10^{-4}$	5									
$10^{-6} - 10^{-5}$	6									
$10^{-7} - 10^{-6}$	7									
$10^{-8} - 10^{-7}$	8									
$<10^{-8}$	9									
		Immediate action needed before further operation								
		Action at next occasion after qualification of analysis for improving system								
		Optional								
		No further action needed								

Fig. 5. Risk potential matrix (Extended HAZOP decision matrix)

Illustrative Example 2

Release of material to atmosphere from vent line or vapour line may disperse safely or has toxic effects or can lead to several outcomes such as flash fire, vapour cloud explosion and fire balls. So documenting consequence class in HAZOP work sheet, the score '8' of the most severe consequence will be documented.

No	Guide word / Process Parameter	Detection/ Safeguards	Possible causes	Conse- quences	FC	Recommended Actions	FC	Resp./ Ref.
				Physical effects: Risk related:				

Plant/P&ID : Equipment : Volume :	Process: Function:	Document: Page : Date :

1- Short cut calculations 2-Dynamic simulation 3-deterministic models 4- Event tree 5- Fault tree
6- Historic data base

2.4.3 Step 3: Safety/risk assessment - Risk potential matrix (HAZOP decision matrix)

Figure 6 shows the risk potential matrix (HAZOP decision matrix) used for order of magnitude ranking of events. The rows of the matrix consider frequency class, while the columns show the consequence class. Each cell in the matrix represents a risk category. For the decision process, the matrix is divided into four risk category levels.

Risk level I --- red area --- scenario in this level is intolerable and immediate action (pant or process modification) is needed to reduce that risk category or more detailed quantified analysis has to be carried out in order to find arguments for wrong preliminary decisions.

Risk level II --- grey area --- scenario in this level is tolerable but not acceptable for long period of time so action at next schedule maintenance is needed to reduce that risk category.

Risk level III --- yellow area --- scenario in this level is acceptable and any action to reduce that risk category is optional.

Risk level IV--- green area --- scenario in this level needs no action.

Risk potential matrix (HAZOP decision matrix) may be also used for:

- Documentation of the status of the plant safety
- Selection and development of optimization proposals
- Importance of improvement
- Documentation of improvement achieved

The application of risk potential matrix (HAZOP decision matrix) in the Extended HAZOP is shown in figure 5. Arrows show the transformation of entries from the Extended HAZOP worksheet to the HAZOP decision matrix. The identity number (ID) of each scenario of the Extended HAZOP worksheet is placed in HAZOP decision matrix. Recommended actions for this scenario will be placed from Extended HAZOP sheet to the bottom of HAZOP decision matrix. First HAZOP decision matrix will shows the existing status and second HAZOP decision matrix shows the improved plant status after recommended actions.

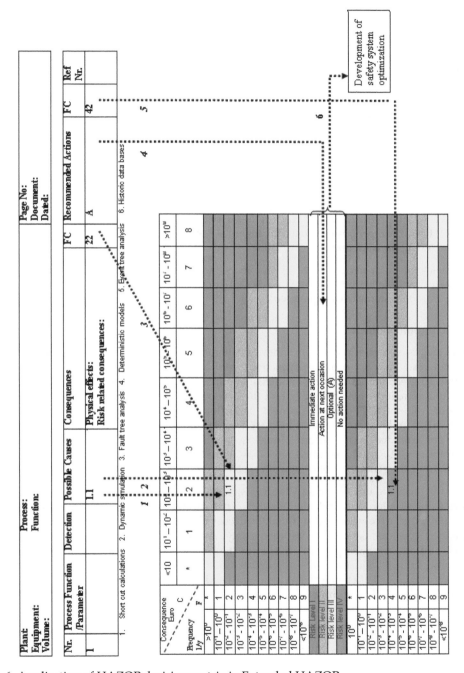

Fig. 6. Application of HAZOP decision matrix in Extended HAZOP

Similarly all results from Extended HAZOP worksheets are transferred to the HAZOP decision matrix. Keeping in view the risk target and depending on the scenario or recommended actions during the Extended HAZOP discussion, analysis team may reach a safety related modification proposal. Next, if safety/ risk optimization is in focus then weak points/scenarios with similar risk are clustered after analyzing HAZOP decision matrix and safety related optimization proposals are developed.

2.4.4 Step 4: Safety/risk system optimization (Development and analysis of optimization proposals)

In this step, safety related optimization proposals are generated and evaluated using dynamic simulation, Event tree analysis and/or Fault tree analysis. The optimization proposals can be developed at two levels:

- Simple optimization proposals e.g. addition of pressure alarm or change of location of sensor within the Extended HAZOP discussion
- Optimization proposals related to severe scenarios by evaluating risk potential matrix (HAZOP decision matrix)

The relevant information such as frequency and damage data will be transferred to economic module for safety related cost calculations and multiobjective decision making (if more than one alternatives developed).

2.5 Economic module

In all stages of design process, economic evaluation is crucial for the evaluation of process alternatives. Various objective functions are available in the literature of chemical engineering economics for economic evaluation of chemical processes. Some quite elegant objective functions, which incorporate the concept of the „time value of money", are net present value (NPV) and discounted cash flow. Business managers, accountants and economists prefer these methods because they are more accurate measures of profitability over an extended time period. However, application of these methods needs certain assumptions[12]. Total annualized cost (TAC) can be used as economic indicators/objective function for the evaluation of design alternatives and economic optimization.

Economic module developed in Visual Basic consists of two distinct sections. First section carries out standard cost calculations (i.e. Fixed capital investment (FC1) and operational cost (OC1)) and compute total annualized cost (TAC1) while second section carries out extended cost calculations i.e. process safety/risk related costs and computes the fixed capital investment related to safety system (FCISS), accident and incident damage related risk cost. Table 3.5 illustrates the difference of cost elements considered in standard practice of cost calculations of chemical process design and in this economic module. Figure 7 shows the simplified block diagram of economic module.

2.5.1 Standard cost calculations

Standard cost calculations involves fixed capital investment (FCI1) and operational cost (OC1). Fixed capital investment (FCI1) includes the cost of design and other engineering and construction supervision, all items of equipment and their installation, all piping,

Standard cost calculations	Economic module used in this work

• FCI1 = Fixed capital investment using either cost equations that have been derived by Ulrich or correlations developed by Guthrie depending on users choice
• OC1= Operating cost (including both direct (e.g. raw material, utilities etc.) and indirect costs (e.g. taxes, overhead cost etc)
• TAC1 = total annualized cost
  = $d \cdot (FCI1) + OC1$
  normally d is taken 0.15-25 but can also be computed using depreciation calculation methods
	• FCI1= Fixed capital investment using either cost equations that have been derived by Ulrich or correlations developed by Guthrie depending on users choice
	• FCI2= FCI1 + fixed capital investment related to safety system (FCISS)
	• OC1= Operating cost (including both direct (e.g. raw material, utilities etc.) and indirect costs (e.g. taxes, overhead cost etc)
	• TAC1 = total annualized cost
	  = $d \cdot (FCI1) + OC1$
	• TAC2 = total annualized cost
	  = $d \cdot (FCI2) + OC1$
	  normally d is taken 0.15-25 but can also be computed using depreciation calculation methods

	Extended cost calculations • RC1 = risk cost 1= Asset risk cost + health risk cost + environmental risk cost • RC2 = risk cost2 = RC1+ production loss risk cost • RC3 = risk cost3 = process interruption cost • TRC = total risk cost = RC2+RC3 • ECC = Extended costs

Table 5. Elements of economic module and difference from standard cost calculations

instrumentation and control systems, buildings and structures, and auxiliary facilities such as utilities, land and civil engineering work. Several capital cost estimate methods ranging from order of magnitude estimate (ratio estimate) to detailed estimate (contractors estimate) are used for the estimation of installed cost of the process units in the chemical plant.

The most commonly used method that provides estimates within 20-30% of actual cost and widely used at design stage involve the usage of cost charts/correlations (Guthrie's article (1969) and book (1974), chapter 5 of Ulrich's 'A guide to chemical engineering process design and economics' (1984), 'Plant design and economics for chemical engineers' by Peters and Timmerhause (1991)) for estimating the purchase cost of major type of process equipment [13-15].

These cost charts / correlations were assembled in the 1960's or earlier and are projected to the date of installation using cost indices or escalation factors such as the chemical engineering plant cost index (published biweekly by chemical engineering magazine),

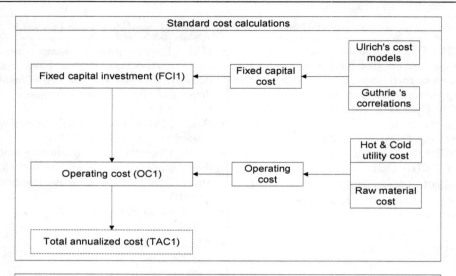

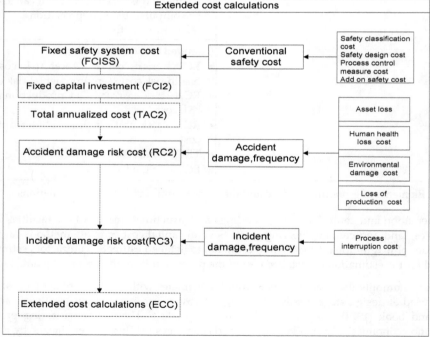

Note:
In Extended cost calculations, the costs such as insurance cost, market loss cost, loss of image and prestige cost should also be considered in addition. But in this module these costs are not included.

Fig. 7. Simplified block diagram of economic module

Marshall and Swift Index (also provided in chemical engineering magazine) and Nelson-Fabaar Index (from the oil and gas journal). For the comparison of process design alternatives, these study estimates for purchased cost of process units using cost charts or equations based on them are adequate. Given the purchase cost of a process unit, the installed cost is obtained by adding the cost of installation using factored-cost methods. For each piece of equipment Guthrie (1969, 1974) provides factors to estimate the direct cost of labor, as well as, indirect costs involved in the installation procedure. The cost elements that are included in the estimation of fixed capital investment are shown in figure 8.

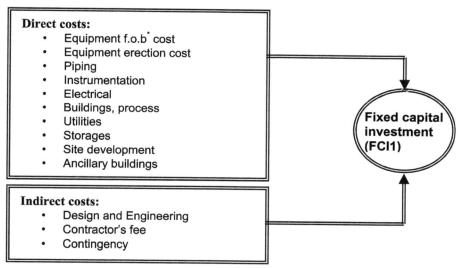

f.o.b ---- freight on board cost i.e. equipment purchase cost plus transport cost

Fig. 8. Typical cost elements for fixed capital investment

The operating cost (OC1) of a chemical plant is divided into two groups:

- Fixed operating cost
- Variable operating cost

The elements in fixed operating cost includes maintenance cost, operating labor cost, laboratory cost, supervision cost, plant overheads, capital charges, taxes, insurance, licence fees and royalty payments while the variable operating cost consists of raw material costs, miscellaneous operating material costs, utilities (services) and shipping and packaging. However this division of operating cost is somewhat arbitrary and depends on the accounting practice of a particular organization. The typical cost elements included in operating cost "OC1" are shown in figure 9.

However, from the existing process optimization point of view, energy cost and raw material costs are more important and often considered.

Economic module developed in this thesis using Visual basic computes fixed capital expenditure using either cost equations that have been derived by Ulrich or correlations

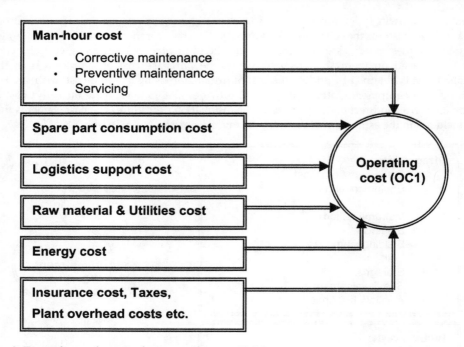

Fig. 9. Typical cost elements for operating cost (OC1)

developed by Guthrie depending on users choice, The significant operating cost for process optimization i.e. energy consumption cost (heating and cooling utilities cost) can also be calculated by using this module. Once the FCI1 and OC1 are calculated, then total annualized cost is obtained using the following equation:

$$\text{Total annualized cost (TAC1)} = d \cdot (\text{FCI1}) + \text{OC1} \qquad (1)$$

Here d is depreciation or capital recovery factor and normally taken between 0.15-0.25 but can also be computed using depreciation calculation methods e.g. double declining balance method.

2.5.2 Extended cost calculations

The second section of the economic module (see Figure 7) carries out Extended cost calculations, which considers the fixed capital investment related to safety system, and risk cost due to accident and incident damage.

i.   Fixed capital investment related to Safety system:

The fixed investment related to safety system is calculated by the following equation:

$$FCISS = C_{SD} + \sum_{i=1}^{n} N_{SE,i} \cdot C_{SE,i} \qquad (2)$$

Here, the first term C_{SD} is cost for safety design (i.e. cost related to safety classification, safety requirements and design specification, detailed design and engineering, factory acceptance test or pre-start up acceptance test and start up and correction). Table 3.6 gives the typical cost elements included in C_{SD} calculations. The second term $\sum_{i=1}^{n} N_{SE,i} \cdot C_{SE,i}$ is the sum of the purchase cost of safety equipment. Here CSE,i is the purchase cost of equipment "i" and NSE,i is the number (count) of that equipment. The costs for these devices are based on the recent detailed survey of available costs from various suppliers conducted by Khan and Annyotte (2004), however in the module the user has the possibility to enter the present market costs.

Safety classification cost e.g SIL determination cost	C_{SIL}
Safety requirements and design specifications (SRS) cost	C_{SRS}
Detailed design and engineering cost	C_{DE}
!Miscellaneous Cost:	C_{ME}
Initial training cost	C_{TC}
Factory acceptance test (FAT)/Installation/Pre-startup acceptance test (PSAT) cost	C_{FAT}
Startup and correction cost	C_{SCC}
$C_{SD} = C_{SIL} + C_{SRS} + C_{DE} + C_{ME} + C_{TC} + C_{FAT} + C_{SCC}$	

!power, wiring, junction boxes, operators interface cost

Table 6. Typical cost elements included in C_{SD} of safety system cost calculations

ii.   Fixed capital investment (FCI2)

Then, extended fixed capital investment is calculated by adding FCISS to FCI1.

$$FCI2 = FCI1 + FCISS \tag{3}$$

iii.   Total annualized cost (TAC2)

So, the extended total annualized cost will be calculated using extended fixed capital investment.

$$TAC2 = d \cdot (FCI2) + OC1 \tag{4}$$

Maintenance and repair cost of safety system should also be included in this calculation. But in this economic module these cost elements are not considered.

iv.   Risk cost 1 (RC1)

Risk cost (RC1), which is the sum of property risk cost due to asset loss (PRC), health risk cost due to human health loss (HRC) and environmental risk cost due to environmental damage (ERC). The relations for calculation of these costs used in the module are:

Property risk cost due to asset lost (PRC) is the cost incurred due to lost of physical assets such as damage to property, loss of equipment due to accident/scenario and calculated by the equation below:

$$PRC = \sum_{i=1}^{n} \dot{F}_{A,i} \cdot A_{D,i} \cdot C_{A,i} \cdot t_{op} + \sum_{j=1}^{n} \dot{F}_{I,j} \cdot C_{D,j} \cdot t_{op} \tag{5}$$

$\dot{F}_{A,i}$ is frequency of occurring the hazardous accident, $A_{D,i}$ is damage area due to that accident, $C_{A,i}$ is the asset cost per unit area, t_{op} is total operation time, $\dot{F}_{I,j}$ is incident occurring frequency and $C_{D,j}$ is incident damage cost.

Health risk cost due to human health lost (HRC) is the cost of fatality and/or injury due to the accident scenario under study.

$$HRC = \sum_{i=1}^{n} \dot{F}_{A,i} \cdot N_{Peop,eff} \cdot C_{H,life} \cdot t_{op} \tag{6}$$

Here, $N_{peop,eff}$ is the number of person affected due to accident and is equal to $N_{peop,eff} = pop \cdot \wp$. Where POP is the population around the area of accident and $\wp$ is the population distribution factor ($\wp$ is 1 if population is uniform distributed (maximum value) and $\wp$ is 0.2 if population is localized and away from the area of accident (minimum value)) and $C_{H,life}$ is dollar value of human life or health. Though attempts to put value on human life have caused criticism and it changes from place to place. But a value for this can be obtained by dividing the annual gross national product by the annual number of births or by estimating how much money the person would have earned if not killed by the accident (Tweeddale 2003). A value for cost of loss of lives, marginal cost to avert the fatality, for the highest category of involuntariness risk 14 x 106 $ is used in this work (Passman, H.J. et al. 2003).

Environmental risk cost due to environmental damage is the cost incurred due to environmental damage.

$$ERC = \sum_{i=1}^{n} \dot{F}_{ED} \cdot A_{ED,i} \cdot C_{ED,i} \cdot t_{op} \tag{7}$$

Where $A_{ED,i}$ is the environmental damage area due to scenario "i", $\dot{F}_{ED}$ is the frequency of release of material to environment and $C_{ED,i}$ is the environmental damage cost per unit area.

so the sum of these three risk costs gives:

$$RC1 = PRC + HRC + ERC \tag{8}$$

v.   Risk cost 2 (RC2)

Risk cost 2 (RC2), which is the sum of risk cost1 (RC1) and production loss risk cost (PLRC), accounts for accident damage risk cost. Here, production loss risk cost due to asset damage (PLRC) accounts for the cost due to the production loss because of accident and given by:

$$PLRC = \sum_{i=1}^{n} \dot{F}_{A,i} \cdot t_d \cdot \dot{C}_p \cdot t_{op} \tag{9}$$

Where PLRC is the production loss risk cost, td is the time lost due to accident and $\dot{C}_p$ is the production loss value in \$/h. Thus Risk cost 2 (RC2) is

$$RC2 = RC1 + PLRC$$

vi.   Risk cost 3 (RC3)

Risk cost 3 (RC3), which is sum of process interruption cost due the spurious trip of the safety system and process interruption cost because of safe shut down to avoid from accident, accounts for incident damage risk cost and calculated as follow:

$$RC3 = (\sum_{i=1}^{n} \dot{F}_S^{trip} \cdot t_{trip} + \dot{F}_R^{trip} \cdot t_{dR}) \cdot \dot{C}_p \cdot t_{op} \tag{10}$$

Here, $\dot{F}_S^{trip}$ is spurious trip frequency, $\dot{F}_R^{trip}$ is safe shut down frequency when trip system demand arises, ttrip is down time due to spurious trip and t_{dR} is down time to safe shut down when trip system demand arises.

vii. Total Risk cost (TRC)

Total Risk cost (TRC) is the sum of all risk costs:

$$TRC = RC2 + RC3 \tag{11}$$

Total risk cost can be annualized by dividing it with total operation time (t_{op}):

$$TAC_{risk} = TRC / t_{op}$$

viii. Extended Cost (ECC)

Extended cost calculations (ECC) is Life cycle related cost and calculated as follow :

$$ECC = FCI2 + PVC \tag{12}$$

Here, PVC is present value of the annual costs (OC1, TAC_{risk}) and calculated as follow:

$$PVC = (OC1 + TAC_{risk} - Insurance \cos t) \cdot \frac{1 - (1 + R)^{-t_{ly}}}{R}$$

R is the present interest rate and t_{ly} is the number of years (predicted life of system).

Besides, the cost elements mentioned above in Extended cost calculation section, the other elements such as warranty/insurance cost, lost of image and prestige cost, market lost cost should also be considered but quantification of these elements is still almost impossible.

2.6 Environment module

Environment module consists of four steps and introduced an environmental performance index (EPI1) for evaluation of environmental performance and environmental pollution index (EPI2) as environmental objective to be integrated along with economics. The environmental performance index (EPI1) is calculated by combining total PEI based on

WAR algorithm[16,17], resource depletion, energy conservation and fugitive emission rates while environmental pollution index (EPI2) is calculated by combining total PEI based on WAR algorithm and fugitive emissions because in this case other factors like resource depletion and energy consumption will be integrated in economic module or objective function. The Analytic hierarchy process (AHP) is used as multicriteria decision analysis tool for combining these different impacts and determination of weighting factors of individual impact categories in total PEI and later on in environmental performance index (EPI1) and environmental pollution index (EPI2) calculations. The module is developed using Microsoft Visual Basic 6.0 and WAR GUI (WAR graphical user interface) is integrated in the user plate form. The steps are:

Step I : Problem definition and data gathering

Step II : Individual impact categories calculation

Step III: Determination of weighting factors

Step IV: Environmental performance index calculation

Figure 10 shows the simplified block diagram of environment module and tasks to be performed.

2.6.1 Step I: Problem definition and data gathering

The primary task in step 1 is problem framing and scope definition. Information such as material and energy balance information, process conditions, process technology and nature of used materials/chemicals should be retrieved from process module. Process flow diagram is to be re-examined for identification of additional waste and emission streams. Collect additional data and information for environment evaluation to fill gaps. As sources of emissions such as fugitive emission sources, venting of equipment, periodic equipment cleaning, incomplete separations etc. are often missing in process so process is analyzed to identify these sources.

2.6.2 Step II: Individual impact categories calculation (Potential environmental impact calculations based on WAR algorithm)

The software WAR GUI (waste reduction algorithm graphical user interface) from the US Environmental Protection Agency is used to calculate individual potential environmental impacts. The generalized formula based on WAR algorithm for calculating individual PEI is given in equation 13.

$$PEI_L = (\dot{M}_b \cdot \sum_{k}^{Comps} x_{kb} \cdot \psi_{kL} + \dot{Q}_r \cdot \psi_L^E) / \dot{M}_p \qquad [Impact/kg\,product] \qquad (13)$$

Where PEI_L is the potential environmental impact of category L, $\dot{M}_b$ is mass flow rate of base (effluent) stream, x_{kb} is the mass fraction of component k in the base stream, ψ_{kL} is the normalized impact score of chemical k for category L, $\dot{Q}_r$ is energy rate supplied for separation and ψ_L^E is the normalized impact score of category L due to energy. The sensitivity analysis results of individual potential environmental impact with respect to optimization variables should also be performed.

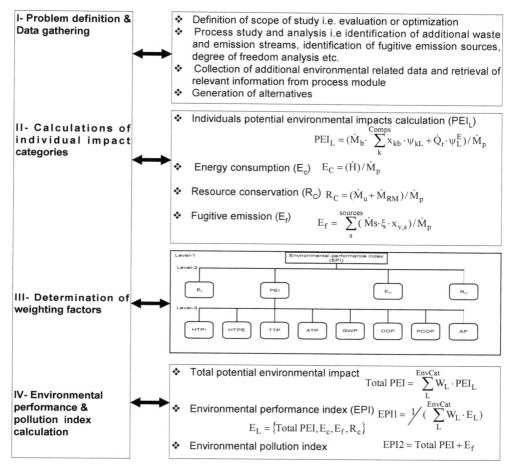

| I- Problem definition & Data gathering | ❖ Definition of scope of study i.e. evaluation or optimization
❖ Process study and analysis i.e identification of additional waste and emission streams, identification of fugitive emission sources, degree of freedom analysis etc.
❖ Collection of additional environmental related data and retrieval of relevant information from process module
❖ Generation of alternatives |

II- Calculations of individual impact categories

❖ Individuals potential environmental impacts calculation (PEI_L)

$$PEI_L = (\dot{M}_b \cdot \sum_{k}^{Comps} x_{kb} \cdot \psi_{kL} + \dot{Q}_r \cdot \psi_L^E)/\dot{M}_p$$

❖ Energy consumption (E_c) $E_C = (\dot{H})/\dot{M}_p$

❖ Resource conservation (R_c) $R_C = (\dot{M}_u + \dot{M}_{RM})/\dot{M}_p$

❖ Fugitive emission (E_f) $E_f = \sum_{s}^{sources} (\dot{M}s \cdot \xi \cdot x_{v,s})/\dot{M}_p$

III- Determination of weighting factors

IV- Environmental performance & pollution index calculation

❖ Total potential environmental impact $\text{Total PEI} = \sum_{L}^{EnvCat} W_L \cdot PEI_L$

❖ Environmental performance index (EPI) $EPI1 = \frac{1}{} (\sum_{L}^{EnvCat} W_L \cdot E_L)$

$E_L = \{\text{Total PEI}, E_c, E_f, R_c\}$

❖ Environmental pollution index $EPI2 = \text{Total PEI} + E_f$

Fig. 10. Simplified block diagram of environment module

Energy consumption factor (E_C)

Energy consumption factor refers the total amount of energy consumed in the process per unit of product and is calculated as follow:

$$E_C = (\dot{H})/\dot{M}_p \qquad [kJ / kg \text{ product}] \qquad (14)$$

Here $\dot{H} = \dot{M}_{steam} \cdot \hat{h}_{steam} + E_E$ where $\dot{M}_{steam}$ is the mass flow rate of steam [kg/h], $\hat{h}_{steam}$ is the enthalpy of steam per kg [KJ/kg], E_E is electrical energy consumed per unit time [KJ / h] and $\dot{M}_p$ is product rate [kg/h].

The sensitivity analysis of this factor with respect to optimization variables should also be performed.

Resource conservation factor (R$_C$)

The resource consumption refers all needed raw materials and utilities used and given by:

$$R_C = (\dot{M}_u + \dot{M}_{RM}) / \dot{M}_p \qquad \text{[kg / kg product]} \qquad (15)$$

Where R_C is the resource conservation factor, $\dot{M}_U$ is utilities consumption rate, $\dot{M}_{RM}$ is raw material consumption rate.

Fugitive emission factor (E$_f$)

Fugitive emissions are unplanned or unmanaged, continuous or intermittent releases from unsealed sources such as storage tank vents, valves, pump seals, flanges, compressors, sampling connections, open ended lines etc and any other non point air emissions. These sources are large in number and difficult to identify. These emission rates depends on factors such as the age and quality of components, specific inspection and maintenance procedures, equipment design and standards of installation, specific process temperatures and pressures, number and type of sources and operational management commitment[18]. However, four basic approaches for estimating emissions from equipment leaks in a specific processing unit, in order of increasing refinement, in use are:

- Average emission factor approach
- Screening ranges approach
- EPA correlation approach
- Unit-specific correlation approach

All these approaches require some data collection, data analysis and/or statistical evaluation. On the other hand, using fundamental design / engineering calculations for accurate fugitive emission estimations for each source present in the process industry are difficult due to:

- large number and type of fugitive emission sources
- dependence of emission rates on other factors along with design and operating conditions e.g. installation standards, inspection and maintenance procedure etc.

As focus in this work is to integrate fuggitve emissions into environmental performance evaluation and optimization objectives so average emission factor approach giving a bit over estimates are used. Average emission factors for estimating fugitive emissions from fugitive sources found in synthetic organic chemical manufacturing industries operations (SOCMI) obtained from the US Environmental Protection Agency L & E Databases are used. The relation used in this work for calculation of fugitive emissions is:

$$E_f = \sum_{s}^{sources} (\dot{M}s \cdot \xi \cdot x_{v,s}) / \dot{M}_p \quad \text{[kg/ kg product]} \qquad (16)$$

Here E_f is fugitive emission factor per unit of product, $\dot{M}_s$ mass flow rate through the source 's', ξ is average emission factor and $x_{v,s}$ is mass fraction of volatile component through source 's' and $\dot{M}_p$ is product rate. It is assumed $x_{v,s}$ for the process fluids through fugitive sources such as pump seals, valves, flanges and connection is equal to 1, i.e. fluids are composed entirely of volatile compounds.

2.6.3 Step III: Determination of weighting factors (Application of multicriteria decision analysis technique)

The integration of these individual impact categories into one index is a hierarchical multicriteria decision analysis problem. The analytic hierarchy process (AHP) is used for this purpose[19] and a computer programme for it is developed in VB 6.0. In this stage, first a hierarchical structure of the problem, which is structured hierarchically similar to a flow chart, is constructed. The overall objective is placed at the top while the criteria and sub-criteria are placed below. For example, as shown in figure 11, the overall objective Environmental performance index (EPI1) is placed at the top (level 1), then below (level 2) are criterias Total PEI, Ef , Ec and RC and after this (level 3) sub-criterias as HTPI, HTPE, TTP, ATP, GWP, ODP, PCOP and AP. After this using the numerical scale given in table 2.6, two pairwise comparison matrices (see Table 7 and 8) are constructed for determination of weights for aggregation of individual impact categories of WAR to total PEI and for determination of weights of total PEI, Ef, Ec and RC to Environmental performance index (EPI1).

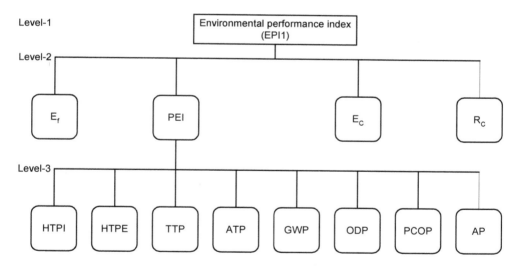

Fig. 11. Hierarchical structuring of multicriteria decision analysis problem for integrating individual environmental impacts

The right hand upper diagonal information in both matrices is to be provided by the decision maker giving the relative importance of the two criteria using the numerical scale of table 2.6 while the left hand lower diagonal is the reciprocal of the right hand upper diagonal. Once these pair wise comparison matrices are constructed, then developed computer programme using the AHP method, determines the weighting of individual impact categories. The level of inconsistency of decision makers input is checked by

Pairwise comparison matrix								
	HTPI	**HTPE**	**TTP**	**ATP**	**GWP**	**ODP**	**PCOP**	**AP**
HTPI	1	A12	A13	A14	A15	A16	A17	A18
HTPE		1	A23	A24	A25	A26	A27	A28
TTP			1	A34	A35	A36	A37	A38
ATP				1	A45	A46	A47	A48
GWP					1	A56	A57	A58
ODP						1	A67	A68
PCOP							1	A78
AP								1
W$_L$	W1	W2	W3	W4	W5	W6	W7	W8

Table 7. Pairwise comparison matrix for individual impact categories at level 3

Pairwise comparison matrix				
	PEI	**R$_C$**	**E$_C$**	**E$_f$**
PEI	1	A12	A13	A14
R$_C$		1	A23	A24
E$_C$			1	A34
E$_f$				1
W$_L$	W1	W2	W3	W4

Table 8. Pairwise comparison matrix for individual impact categories at level 2

consistency ratio before giving the output. Consistency ratio less than 0.1 is good and for ratios greater than 0.1, the input to pair wise matrix should be re-evaluated.

2.6.4 Step IV: Environmental performance & pollution index calculation

In the final step, first Total PEI is determined by multiplying each impact category values with its relevant weighting factor WL as given below:

$$Total\ PEI = \sum_{L}^{EnvCat} W_L \cdot PEI_L \tag{17}$$

After calculating Total PEI, Environmental performance index (EPI1) is determined for each alternative by multiplying the values of Total PEI, Ef, EC and RC with its relevant weighting factor WL (table 8) as given below:

$$EPI1 = \frac{1}{\left(\sum_{L}^{EnvCat} W_L \cdot E_L \right)} \tag{18}$$

Where $E_L = \{Total\ PEI, E_c, E_f, R_c\}$

and environmental pollution index (EPI2) is calculated as follow

$$EPI2 = Total\ PEI + E_f \tag{19}$$

The higher value of environmental performance index (EPI1) shows that the process is environmentally better and vice versa. While the higher value of environmental pollution index (EPI2) shows that the environmental performance of process is worse.

2.7 Data manager

The relevant information generated from process module, safety module and environment module for each alternative is transferred to data manager. This information is used to formulate process diagnostic tables and multiobjective decision-making problem formulation. These tables consist of mass input/output table, energy input/output table, capital and utility annual expense summary, environmental impact summary and frequency of occurance of an event and their consequence categories and safety cost.

2.8 Stage III: Multiobjective decision analysis/ optimization

The purpose of this layer/stage is to set up multiobjective decision making/optimization among these conflicting objectives. The aim is to find out the trade-off surface for each alternative and /or complete ranking of alternatives. The calculation loop used for it is shown in figure 12.

In each independent performance module i.e. economic, environment and safety module, relevant information is generated and transferred automatically or manually to data manager for each alternative generated or under study. Before transferring the values of performance objective functions, each objective function is optimized within their independent module such as:

- Process/Economic optimization of each alternative is carried out using SQP optimization algorithm build within Aspen Plus™.
- The lower and upper limits for Environmental objective functions are calculated using environmental module from the material and energy balance information from process model.
- Safety/risk aspects are optimized in the safety module and information such as hazard occurance frequency, safety cost data (fixed safety system cost, accident and/or incident damage risk cost) for each alternative is transferred to the data manger.

Depending on the case under study or objectives of the study, graphical tool box of MatLab and/or multiobjective optimization technique (goal programming) or multiattribute decision analysis technique (PROMETHEE and/or AHP) is used for multiobjective decision analysis. The Data Manager is linked with MatLab 7.0 via Excel link Toolbox in the

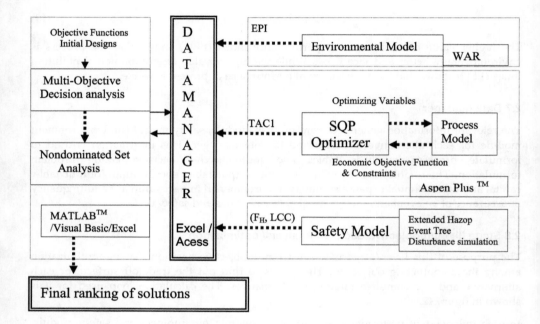

Fig. 12. Calculation loop for multiobjective optimization

integrated interface. Aspen plusTM is linked in the integrated interface via Visual basic 6.0 and Microsoft Excel. AHP technique is also programmed for the cases under study in Visual basic 6.0. The computer realization of these links in the integrated interface is explained in chapter four.

2.9 Stage IV: Design evaluation

The purpose of this layer/stage is to select the best alternative and/or find the complete ranking of alternatives under study based on the results of third stage/layer of the developed methodology. Pareto approach (non dominated analysis) or PROMETHEE is used for this purpose.

3. References

[1] Ramzan, N., Naveed, S., Feroze, N. and Witt, W.; "Multicriteria decision analysis for safety and economic achievement using PROMETHEE: A case study" Process Safety Progress (A Journal of Americal Institute of Chemical Engineering), 28(1), 68-83 (2009).

[2] Ramzan, N., Degenkolbe, S. and Witt, W.; "Evaluating and improving environmental performance of HC's recovery system: A case study of distillation unit", Chemical Engineering Journal (Journal published by Elsevier), 40(1-3), 201-213(2008). ISSN: 1385-8947.

[3] Ramzan, N., Compart, F. and Witt, W.; "Methodology for the generation and evaluation of safety system alternatives based on extended Hazop" Process Safety Progress (A Journal of Americal Institute of Chemical Engineering), 26(1),35-42 (2007).

[4] Ramzan, N., Compart, F. and Witt, W.; "Application of extended Hazop and event tree analysis for investigating operational failures and safety optimization of distillation column unit" Process Safety Progress (A Journal of Americal Institute of Chemical Engineering), 26(3),248-257 (2007).

[5] Ramzan, N., Witt, W.; "Multiobjective optimization in distillation unit: A case study", The Canadian Journal of Chemical Engineering, 84(5), 604-613(2006).

[6] Seader et al. (1999), Process design principles-Synthesis, Analysis, and Evaluation, John Wiley & Sons Inc. New York, 338-370.

[7] Nawaz, Z. Mahmood, Z. (2006) Importance Modeling, Simulation and Optimization in Chemical process Design, The Pakistan Engineer (Journal of Institute of Engineers Pakistan), 38-39.

[8] Cameron, I., Raman, R. (2005), Process Systems Risk Management, Elsevier Academic Press, NY, ISBN 0-12-156932-2

[9] Crowl, D.A., Louvar, J.F. (1999), Chemical Process Safety: Fundamentals with applications," Prentice Hall, New York.

[10] Kletz, T.A. (1997), Hazop-past and future, Reliability engineering and system safety, 55, 263-266.

[11] Lees, F.P. (1996), Loss prevention in CPI, Butterworth's, London, UK.

[12] Peters, M. S. and K.D. Timmerhaus (1991), Plant Design and Economics for Chemical Engineers, Ed.2nd, McGraw-Hill, New York, 90-145.

[13] Douglas, J. (1998), Conceptual design of chemical processes, McGraw Hill Inc.

[14] Guthrie, K.M. (1969), Data and techniques for preliminary capital cost estimating, Chem. Eng., 114-1421.

[15] Guthrie, K.M. (1974), Process plant estimating, evaluation and control, Craftsman, Solano Beach, CA.

[16] Cabezas, H. Bare, C. & Mallick, K. (1999), Pollution prevention with chemical process simulators: the generalized waste reduction (WAR) algorithm-full version. Computers and Chemical Engineering, 23,623-634.

[17] Cabezas, H. Bare, C. & Mallick, K. (1997), Pollution prevention with chemical process simulators: the generalized waste reduction (WAR) algorithm, Computers and Chemical Engineering, 21s, s305-s310.

[18] Dimian, A. C. (2003), Integrated design and simulation of chemical processes, 1st Eds., Elsevier Netherlands, 1-30,113-134.

[19] Dev, P.K. (2004), Analytic hierarchy process helps evaluate project in Indian oil pipelines industry, International journal of operation and production management, 24(6), 588-604

3

Optimization of Spouted Bed Scale-Up by Square-Based Multiple Unit Design

Giorgio Rovero, Massimo Curti and Giuliano Cavaglià
Politecnico di Torino
K&E Srl
Italy

1. Introduction

Among several configurations typical of gas-solids fluidization, spouted beds have demonstrated to be characterized by a number of advantages, namely a reduced pressure drop, a relatively lower gas flow rate, the possibility of handling particles coarser than the ones treated by bubbling fluidized beds. Additionally, significant segregation is prevented by the peculiar hydraulic structure.

Spouted beds appear to go through a revival, testified by a very recent and comprehensive book on the topic (Epstein & Grace, 2011). This renewed interest arises by implementing new concepts in scaling-up spouting contactors and devising potential applications to high temperature processes, noticeable examples being given by pyrolysis and gasification of biomass, kinetically controlled drying of moist seeds to guarantee the requested qualities and polymer upgrading processes.

2. Generalities on fluidization

Fluidization is a hydrodynamical regime in which a bed of solid particles is expanded and suspended by an upward fluid flow. This regime is established when the fluid velocity reaches a value corresponding to the minimum fluidization. The basic design of a fluidized unit is carried out by considering a vessel having a cross section of any shape (circular, squared or rectangular) with a perforated bottom which separates the volume holding the solids from the lower gas plenum.

Fluidized beds show a number of features which are summarized below:

- forces are in balance and there is no net force acting in the system;
- the solid particle bulk exhibits a liquid-like behaviour: the surface of the solids remains horizontal by tilting the vessel;
- if two or more vessels operating in a fluidization regime are connected, the solids reach an identical hydrostatic level;
- in the presence of a side opening under the bed surface, particles gush as a liquid flow;
- heterogeneous bodies may float or sink, depending on their actual density.

Fluidization, besides being influenced by the solid characteristics, depends on the physical properties of the fluid and its superficial velocity. When this parameter is very low, the fluid merely percolates through the particles and no movement is induced, this condition being defined as "static bed". By rising the flow rate, frictional forces between particles and fluid increases: when the upward component of force counterbalances the particle weight, the minimum condition to expand the bed is reached. When all the particles are suspended by the fluid, the bed can be considered in a state of "incipient fluidization" and the pressure drop through any bed section equalizes the weight of the fluid and solids in that section. By further increasing the velocity, some phenomena of instability such as "bubbling or turbulent fluidization" may occur, depending on the system geometry and particle properties. In a gas-solid system operated at high fluid velocity, gas bubbles tend to coalesce and grow in volume during their upward travel; if the bed is not wide enough, a gas bubble can take all the vessel cross section, then the solid particles are lifted as a piston, giving origin to the so-called "flat slugging". This undesired occurrence easily happens with coarse particles as well with cohesive powders. Finally, when a critical value is reached, the velocity of the gas is high enough to transport individually or in clusters the bed particles in a "pneumatic conveying" fashion. These hydraulic regimes are schematically shown in Figure 1.

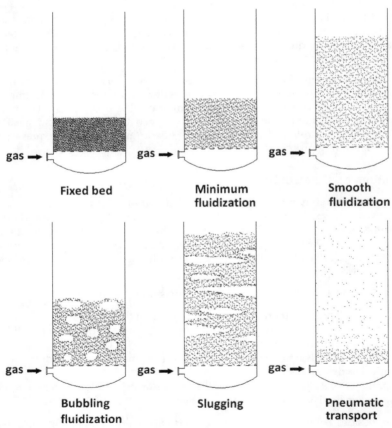

Fig. 1. Schematic representation of various fluidization regimes

In addition to properties of the fluidization medium, particle features play an essential role. A simple mapping was proposed (Geldart, 1973) to group particulate solid materials in four well-defined classes according to their hydrodynamic behaviour. To follow this categorization, Figure 2 shows the four different regions, proper of air/solids systems at ambient fluidization conditions and particles of the same shape, excluding pneumatic transport conditions.

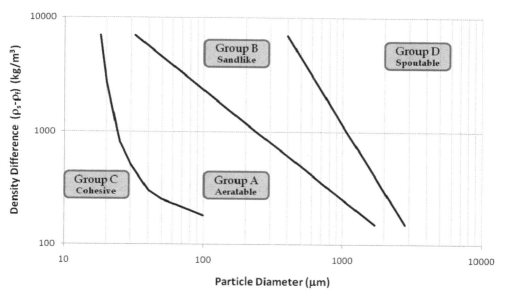

Fig. 2. Particle classification according to Geldart

The particles of the regions can be described as follows:

A type or "aeratable": solids with small diameter and density lower than about 1400 kg/m³. The minimum fluidization velocity can be reached smoothly; then fine bubble fluidization occurs at higher gas velocities.

B type or "sandlike": these particles are coarser than the previous ones, ranging from 40 to 1000 µm and densities from 1400 to 4000 kg/m³. A vigorous fluidization with large bubbles may be established.

C type or "cohesive": very fine powders, with a mean diameter generally lower than 50 µm. Strong interparticle forces render fluidization difficult.

D type or "spoutable": coarsest particles within a broad density range. These systems are characterized by a high permeability, which generates severe channelling and uneven gas distribution: the standard fluidization geometry should be modified to give origin to spouted systems.

3. Fluidization versus spouting

Fluidization is an operation characterized by several interesting peculiarities with desirable associated to non-optimal features. On one hand this technique guarantees a smooth and liquid-like flow of solid particulate materials that allows continuous and easily controlled operations. The good mixing of solids provides a large thermal flywheel and secures isothermal conditions throughout the reactor. If some minor wall-effects are neglected, the solids-to-fluid relationship is independent of the vessel size, so this operation can be easily scaled up and large size operations are possible.

On the other hand, fluidization may reach conditions of instability, such as bubbling or slugging, which usually represent a situation of inefficient contact between the two phases. Moreover, in high temperature operations or with sticky particles, solids sintering or agglomeration easily occur. Finally, but not less important, fines generation, erosion of vessel, internals and pipes is a serious problem caused by the random and intense movement of particles (i.e. particle or carry over exerted by the fluid.

Mass and heat transfer related to physical and chemical reactions are kinetically limited by surface area of large particle, either when the operation occurs in the fluid phase or on the solids. In these cases fluidized systems must operate with fine particulate materials; examples are given by heterogeneous catalysis or combustion/gasification of coal fines.

For reasons intrinsic to many processes (agricultural products upgrading, agglomeration, pelletization, etc.), large particle handling is required and fluidization does not represent an optimal technology. Material comminution to reach the size required by conventional fluidization is an additional negative aspect which increases the exergetic overall process cost. A very noticeable gas rate is required to reach fluidization of large particles, which often far exceeds the amount required for the physical or chemical operation considered. It should also be noted that fluidized systems are operated at a gas rate double or triple with respect to the minimum fluidization velocity to confer the system adequate mixing and avoid any dead zone. In conclusion, fluidization appears an interesting operation thanks to an easy scale-up, though its extensive feature (large gas flow rate need) as well as its intensive characteristics (random fluid-to-particle hydraulic interaction) counterbalance its favourable aspects to some extent.

A spouted bed can be realized by replacing the perforated plate distributor typical of a standard fluidized bed with a simple orifice, either located in the central position of a flat bottom or at the apex of a bottom cone, whose profile helps the solids circulation and avoids stagnant zones. Examples of non-axial orifices appear in the scientific literature, too. The fluidizing gas enters the system at a high velocity, generates a cavity which protrudes upward through the "spout", which, having an almost cylindrical shape, can be characterized by its spout diameter value D_s. When the gas flow rate is large enough, the spout reaches the bed surface and forms a "fountain" of particles in the freeboard. The fountain can be more or less developed depending on the gas rate and the overall system features. After falling on the bed surface, the solids continue their downward travel in the "annulus" surrounding the spout and reach different depths before being recaptured into the spout. The dual hydrodynamics, good mixing in spout and fountain and piston flow in the peripheral annulus with alternated high and slow interphase transfer, in the spout and in the annulus respectively, makes spouted beds unique reactors.

Consequently, spouted beds offer very peculiar features, which can be summarized as:

• 	very regular circulation of particles and absence of dead zones;

- reduced pressure drop and lower gas flow rate required to attain solids motion with respect to the minimum fluidization velocity, this result being possible as the gas transfers its momentum to a limited portion of solids constituting the whole bed;
- wide range of operating conditions starting from a value slightly exceeding the minimum spouting velocity ;
- possibility of handling coarse particles having a wide size range and morphology.

4. Spouted beds

The term "Spouted Bed" was coined with an early work carried out at the National Research Council of Canada by (Gishler & Mathur, 1954). A comprehensive book by (Mathur & Epstein, 1974) provided a systematic summary of the scientific work done in the years. Very recently (Epstein & Grace, 2011) the most advanced knowledge in the field has been updated.

Spouted beds were originally developed as an alternative method of drying moist seeds needing a prompt and effective processing. Sooner the interest in spouted beds grew and their application included coal gasification and combustion, pyrolysis of coal and oil shale, solid blending, nuclear particle coating, cooling and granulation as well as polymer crystallization and solid state polymerization processes. Fundamental studies were carried out to establish design correlations, the performance of spouted beds as chemical reactors, motion patterns and segregation of solids, gas distribution within the complex hydrodynamics.

Some additional improvement in the gas-to-solids contacting can be provided by independently aerating the annulus, thus generating the so-called "Spout-fluid Beds" (Chatterjee, 1970). Again, a perforated draft-tube can be placed to surround the spout, thus contributing in terms of stability and operational flexibility (Grbavčić et al., 1982).

Most studies were carried out in plain cylindrical geometries, either full sectional, half sectional or even in a reduced angular section of a cylinder in order to explore scale-up possibilities. In any case adding a flat transparent wall has been demonstrated to interfere to a moderate extent with the solids trajectory vectors within the annulus (Rovero et al., 1985) while does not affect the measurement of the fundamental parameters of spouted beds (U_{ms}, H_m, D_s). A limited number of examples consider multiple spouting, either in parallel or in series, squared and rectangular cross sections (Mathur & Epstein, 1974).

5. Design bases

A typical spouted bed scheme is given in Figure 3. This representation depicts the gas inlet and outlet, the upward movement of solids in the spout, their trajectories in the fountain and the subsequent descent in the annulus to reach the spout again at depths which depends on the path-lines originated by the landing position on the bed surface. The particle holdup can be loaded batchwise or, alternatively, in a continuous mode. The latter option depends on process requirements, nevertheless a continuous solids renewal in no way alters the above features. A proper solids feeding should minimize bypass towards the discharge port; in this view a direct feeding over the bed surface appears the best option to guarantee at least one circulation loop in the annulus to all the particles. A direct distributed feed over the fountain is required only in case of particles with high tendency to stick.

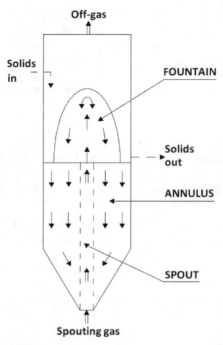

Fig. 3. Schematic of a spouted bed

The solids discharge from continuous operations is generally carried out with an overflow port, unless special process control is required. This case may be given by coating processes, where the total bed surface area should be controlled. In this case a submerged port preferentially discharges coarse material, due to local segregation mechanisms (Piccinini, 1980), while the entire spouted bed retains good mixing capacity, which can be regulated by the fountain action.

The gas flow distribution between spout and annulus is completely independent whether the solids are batch or continuously fed: part of the gas progressively percolates from the spout into the annulus by moving toward higher elevations in the bed of particles. In case of a bed of sufficient height the gas in the annulus may reach a superficial velocity close to the minimum fluidization velocity of the solids. In this event the annulus is prone to collapse into the spout, thus defining the "maximum spoutable bed depth, or H_m". This parameter represents one of the fundamental criteria to design a spouted bed unit (Mathur & Epstein, 1974); H_m depends on vessel geometry, fluid and particle properties.

The second fundamental parameter is given by the minimum rate of gas required to maintain the system spouting, the so-called "minimum spouting velocity, or U_{ms}". This operating factor can be either determined by an experimental procedure (as described in Figure 10) or can be calculated by the existing correlations.

A spouted bed, thanks to its flexibility, can be operated with a wide range of solids load to fill part or all the cone ("conical spouted beds"), or otherwise to engage also the upper portion of the vessel ("cylindrical spouted beds"). In both cases the conical included angle is in the range of 60 to 90°; by further diminishing the angle instability in solid circulation might occur, while increasing excessively the angle decreases solids circulation at the base. A gross criterion that distinguishes conical and cylindrical spouted beds can be given considering the type of reaction to carry out: when solid phase undergoes a fast surface transformation, the optimum residence time of the gaseous phase is very short. This condition is satisfied by shallow beds as in the case of catalytic polymerization or coal gasification and pyrolysis. When the reaction is controlled by heat or mass transfer, the gas-solid contact must be adjusted with a deeper bed configuration as in drying, coating, solid phase polymerization, etc..

Thanks to this flexibility, the mean residence time of the solids in continuous operations can be regulated by optimizing the solids hold-up in a single vessel, which gives origin to a well mixed unit, or otherwise it is possible to conceive a cascade of several units to have a system approaching a plug flow. In the latter case square based units can have a number of advantages over a conventional cylindrical geometry. Specifically, the construction is cheaper, more compact and the heat dissipation toward the outside lower. Some scientific aspects remain open though: the design correlations that should be validated, the stability of multiple units proved both during the start-up and at steady state conditions and the possibility of fully predicting the solids residence time distribution as a function of geometry and number of stages.

6. Spouting regime

Stable spouting can be obtained by satisfying two hydrodynamic requirements: 1) the bed depth must be lower than the H_m value and 2) the gas flow rate has to exceed U_{ms}. From an initial condition of a static bed with a nil gas flow, by increasing the gas flow a certain pressure drop is built up through the bed of particles. The graph given in Figure 4 describes this hydrodynamic evolution, which implies a pressure drop/flow rate hysteresis between an increasing flow and the reverse situation. The hysteresis is caused by different packing conditions of the bed particles, that expand to attain a loose state once a spouting condition is reached. Starting from the static bed condition denoted by A, the pressure drop increases with the fluid velocity and reaches a maximum pressure drop (ΔP_M at B). With an additional increase of the gas velocity, the bed displays a moderate progressive expansion and a corresponding decrease of pressure drop to reach C. Finally an abrupt spouting leads to a sudden decrease of pressure drop which stabilizes at an nearly constant value (D), which is maintained in all the operating range of gas rate. This situation represents a stable spouting. In case of fluid velocity decrease, the pressure drop remains constant down to the spout collapse (E), which compacts the system to some extent and the pressure drop increases again to F, giving origin to the afore mentioned hysteresis. The minimum spouting velocity is recorded at E.

The whole system hydrodynamics is given by knowing ΔP_M, U_{ms}, ΔP_s and the U/U_{ms} ratio chosen for a stable spouting. A recent paper has compared data obtained in the mentioned 0.35 m side square-base unit to the existing literature correlations (Beltramo et al., 2009). The design data for the blower are ΔP_s and U, while the maximum pressure drop and the relative transitory flow rate can be easily generated by a side capacitive device, to be used at the start-up only. It is important knowing that the hydrodynamic transient can be as short as a few seconds, so that the capacitive device can be designed with a characteristic time shorter than a fraction of a minute. In this view, the timing of a spouting process onset is quantified in Figure 15.

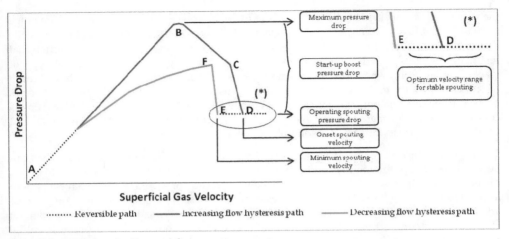

Fig. 4. Hydrodynamic diagram for spouting onset

Figure 5 displays a sequence of pictures that show the spout onset from the initial cavity generation (A) to the full spouting. The third picture qualitatively corresponds to the point B in Figure 4, the fourth picture shows the rapid sequence between C and D, while the last picture may describe any point in the interval E-D, or over.

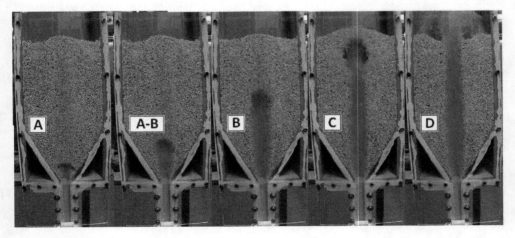

Fig. 5. Photographic sequence of the evolution of a spouting process performed in a squared-based half sectional 0.2 m side unit

7. Scale-up

Due to the peculiarities of spouted beds, their application to industrial processes requires a sound experience and a clear vision of their hydrodynamics since scale-up from laboratory experience is required. A summary of industrial implementations appears in the recent book on spouted beds by (Epstein & Grace, 2011) together with general criteria, though the issue remains open.

Scaling-up a spouted and spout-fluid beds can be tackled according two approaches:

1. increasing the size of a single unit, or
2. repeating side by side several units.

Both routes must be discussed in terms of advantages and drawbacks. The first approach implies a simple geometry and mechanic construction: some doubts arise on the validity of the existing correlations and the overall hydrodynamics in the unit (gas distribution between spout and annulus, solids circulation, etc.). The use of the existing correlations up to a unit diameter (D_c) of about 0.6 m is generally thought fully safe. If a continuous operation is considered, this arrangement gives the solid particulate material a well-mixed behaviour with a broad distribution of particle residence time at the exit of the unit.

Conversely, if a sequence of multiple beds is realized, achieving a fully independence of the units becomes the fundamental goal. In other words a non-interfering system must be designed, so that it is up to the operator decide which unit to start-up or shut-down first according to process needs. Due to the complexity of a multiple system each unit must mandatorily replicate the foreseen behaviour of the basic component. In this case, the residence time distribution of the solids approaches closely a plug flow to meet most process requirements.

Design geometry and regulation criteria are thoroughly discussed in the continuation of this chapter. The design and the construction of a multiple unit implies a careful geometrical optimization to minimize heat loss, investment and operating costs, assure a straightforward start-up, guarantee stability and process performance. According to this key requirement, squared-based units could replace a standard cylindrical section geometry, according to an account presented in literature (Beltramo et al., 2009). A correlation between cylindrical and square-based units is also needed.

8. Experimentation on single and multiple square-based units

This chapter describes a systematic experimentation in a 0.13, 0.20 and 0.35 m side units to correlate the hydrodynamics of square-based spouted beds to the one of a corresponding cylindrical units and define the optimal geometrical configuration to assure solids circulation and transfer to downstream modules when multiple spouted beds in series are considered. The square-based experimental modules were made of wood with a frontal Perspex wall or of AISI 316 SS with a tempered glass window, depending on whether the apparatus had to be operated at room or higher temperatures. A 0.15 m ID cylindrical unit was also used for data comparison. Table 1 provides more geometrical details of the spouted beds. All bases, either frustum shaped or conical were characterized by an included angle of 60°. The spout orifice extended up of 1 mm over the base to improve solids circulation at the bed bottom, according to suggestion existing in the literature. The vessels were 1.5 m or 2 m high to allow the measurement of the maximum spoutable bed depth for all the solids tested.

The tests were carried out with several materials to cover a sufficient range of parameters, as they appear in Table 2.

Module type:	Section	Side (m)	Diameter (m)	Equivalent diameter (m)	Height of the unit (m)	Cone / Pyramid angle (deg)
L-13	squared	0.13	-	0.15	1.50	60
D-15	cylindrical	-	0.15	0.15	1.50	60
L-20	squared	0.20	-	0.23	2.00	60
L-35	squared	0.35	-	0.40	2.50	60
half L-20	rectangular	0.20	-	-	2.00	60

Table 1. Geometrical characteristics of spouted test apparatuses

Material	Equivalent mean diameter (mm)	Density (kg/m³)	Sphericity	Repose angle (deg)
PET chips	3.04	1336	0.87	35
Turnip seeds	1.50	1081	1.00	27
Corn	7.82	1186	0.80	27
Soya beans	7.23	1144	0.99	29
Sunflower seeds	6.16	696	0.87	37

Table 2. Physical properties of the particulate material used

Fig. 6. Materials used in the tests

The experimental equipment was composed of several units, whose pictures are given in Figure 7A and 7B. Depending on the flow rate required, air as spouting medium was either provided by two volumetric compressors (total flow rate of about 250 Nm³/hr) or from a blower (flow rate of 350 Nm³/hr). The air flow was cooled through a corrugated pipe heat exchanger to guarantee spouting at a constant room temperature. Two rotameters were used to meter the flow together with gauge pressure recording.

The experimental strategy was directed both to run batch experiments in a single vessel to assess the fundamental spouting parameters (H_m, U_{ms}, a stable U/U_{ms} ratio), as well as

Fig. 7A. See from left to right: a) 0.15 m ID cylindrical Perspex spouted bed vessel, b) 0.13 m side square-based wooden unit; equivalent to a), c) 0.20 m side square-based wooden unit and d) half-sectional 0.20x0.10 m^2 square-based wooden unit.

Fig. 7B. 0.35 m side square-based AISI 316 SS unit with tempered glass frontal window.

optimize the geometrical spouted bed features (details of the base, orifice diameter, particle traps at the gas exit, etc.). Additionally, cylindrical and square-based vessels were comparatively tested.

Continuously operating experiments were directed to design the geometry of internals required to guarantee easy start-up, spouting stability in a multiple stage unit, discharge facilities to minimize "off-spec" products, adequate solids transfer from the feeding port, through the inter-stage weir, to the final overflow discharge.

Figure 8A provides the schematic view of the spouted bed assemblage of D-15, L-13, L-20, half-L-20 and L-35 units for batchwise measurements of the fundamental operating

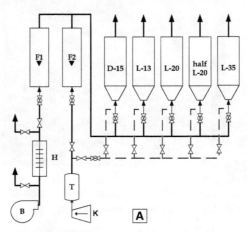

Fig. 8A. Schematic of the spouted bed assemblage for batch measurements of operating parameters (see legend in the below Fig. 8B)

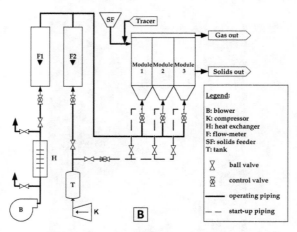

Fig. 8B. Schematic of the three module spouted bed for continuous hydrodynamic measurements

parameters characteristic (H_m, U_{ms}, D_s , mean particle velocity in the annulus, volume of spout and fountain). Figure 8B shows the scheme of the three cell equipment used to optimize geometry of internals, overall structure, effect of continuous solids feeding and residence time distribution of the solid phase.

8.1 Batch unit tests

The hydrodynamic behaviour of a square-based spouted beds was evaluated by exploring a wide range of conditions, from a bed depth corresponding to the frustum height to the maximum spoutable bed depth by operating with all the material shown in Figure 6. A reasonable ample scale-up factor (in excess of 7) was considered by running four square-based units of 0.13, 0.20 and 0.35 m side. Moreover the 0.13 m side unit was compared to the equivalent 0.15 m ID cylindrical unit to identify any difference in terms of U_{ms} and H_m and then validate the applicability of the existing correlations to the non-standard square-based geometry.

8.1.1 Maximum spoutable bed depth

The maximum spoutable bed depth is measured by progressively adding solid granulated material in the vessel and verifying that a stable spout could be formed. Some difficult transition may be encountered at $H \approx H_m$ to make the spouting process neatly evolve from an internal spout to an external well-formed one; in these cases some subjective uncertainty may be left in the measurements. In this case visualization in half-column can be useful (see the photographs of Fig.14).

Table 3 presents the experimental results obtained in two equivalent units, namely the cylindrical 0.15 m ID column (D-15) and the 0.13 m side square-based vessel (L-13); additionally, the larger L-20 vessel provided data useful for scale-up and validation of existing correlations. The test results show good and consistent agreement with the predictions given by literature equations (Malek & Lu, 1965; McNab & Bridgwater, 1977).

	D-15	L-13	L-20
Material	H_m, m	H_m, m	H_m, m
PET chips	0.69	0.65	1.26
Turnip seeds	0.84	0.85	1.44
Corn	0.41	0.46	0.98
Soya beans	0.37	0.36	0.65
Sunflower seeds	0.68	0.69	1.48

Table 3. Experimental values of the maximum spoutable bed depth H_m in the 0.13 m side square-based unit (L-13), in the 0.20 m side square-based unit (L-20) and in the 0.15 m ID cylindrical spouted bed (D-15).

Another interesting result derived from the comparison of the D-15 and L-13 units; the maximum spoutable bed depth values are very close for the same particles and identical orifice, so it allows us to assume that this fluid dynamic parameter does not depend on the cross section geometry. For this reason several correlations predicting H_m can be indifferently applied to both geometries. The herein below Malek-Lu (derived in SI units) and McNab-Bridgwater equations, among others, were compared with the experimental results, as it appears in Figure 9. The agreement can be defined fully satisfactory.

Malek-Lu equation:

$$\frac{H_M}{D_C} = 418 \cdot \left(\frac{D_C}{d_p}\right)^{0.75} \cdot \left(\frac{D_C}{d_i}\right)^{0.40} \cdot \left(\frac{\lambda^2}{\rho_s^{1.2}}\right) \tag{1}$$

McNab-Bridgwater equation:

$$H_M = \frac{D_C^2}{d_p} \cdot \left(\frac{D_C}{d_i}\right)^{2/3} \cdot \frac{700}{Ar}\left(\sqrt{1 + 35.9 \cdot 10^{-6} \cdot Ar} - 1\right)^2 \tag{2}$$

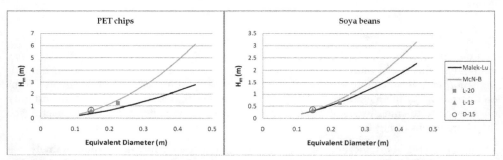

Fig. 9. Comparison of experimental data obtained in square-based and cylindrical units with literature correlations

8.1.2 Minimum spouting velocity

As given in the hydrodynamic diagram of Figure 4, the minimum spouting velocity U_{ms} represents the superficial velocity in the vessel below which spouting does not occur. Aiming to work in a stable situation, good practice suggests to moderately exceed this value by defining an operating spouting regime given by U/U_{ms} > about 1.05, which is the measure of a very modest excess of gas with respect to the minimum spouting condition. It is worthwhile noting that this value can be further reduced if the spouted bed approaches its maximum spoutable bed depth H_m, as the below Figure 10 diagram indicates. The onset spouting and minimum spouting velocities approach as H is closer to H_m. This occurrence also indicates that passing from a submerged to an external spouting is progressively easier as H_m is approached, since the bed of particles is already highly expanded by a very high gas rate (pictures on Figure 14 represent this case). Consistently, these operating conditions generate hydrodynamic diagrams with a much less pronounced hysteresis, as given in a recent paper (Beltramo et al., 2009). To predict the minimum spouting velocity in multiple cell systems an empirical correlation was proposed (Murthy and Singh, 1994). By observing carefully the type of plots appearing in Figure 10, it is possible to note that the maximum bed depth can be inferred from the overlapping of the two curves. This remarkable experimental statement (based on hydrodynamic data) may offer an interesting alternative

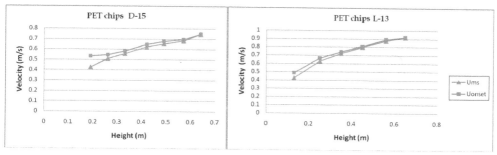

Fig. 10. Onset and minimum spouting velocities vs. bed depth

to direct observations, which are ambiguous in some instances due to bed instability as H_m is approached. Moreover careful extrapolation of the U_{ms} and U_{onset} curves to detect their intersection is a possible method to estimate H_m when the experimental conditions are not suitable for a complete direct measurement.

8.2 Continuously operating multiple units

A continuously operating unit was tested to devise a guideline to design, start-up, gain in stability and proper hydrodynamics in multiple square-based spouted beds.

A picture of the rig is present in Figure 11A, while a 3D scheme of the same unit is shown below, see Figure 11B.

Fig. 11A. Picture of the three-module experimental rig.

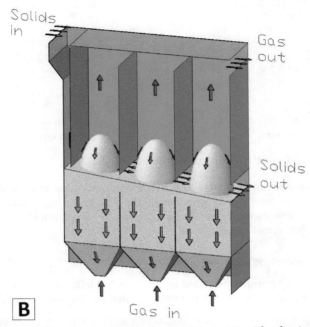

Fig. 11B. 3D diagram of the same continuously operating unit with a horizontal layout of stages

Designing a multiple spouted bed system does not represent a "black art", though may be more complex than other gas/solids contactors. In particular the following design aspects were focused in this project:

- diverting baffle at solids inlet,
- fountain height regulators,
- freeboard baffles between units,
- submerged baffles between units,
- overall layout of a multiple spouting unit

The following paragraphs illustrate the progressive tuning of a multiple stage spouted bed up to achieve safe know-how and run continuous operations.

8.2.1 Solids inlet design

Solid particles were stored in an elevated drum, from which they could flow by gravity, being metered by a rotary valve, a screw feeder or a simple calibrated orifice. The particles were distributed along one entire side of the spouted bed, addressed down by a vertical baffle protruding to a small distance from the bed free surface. This simple device avoided any solids bypass promoted by the fountain. Residence time distribution measurements quantified a bypass as high as about 20% of the total flow rate, in case of baffle absence. The elevation of the baffle over the bed may require some regulation, depending on solids feed rate.

8.2.2 Fountain height regulators

A spouted bed fountain can be defined underdeveloped, developed or overdeveloped, depending on whether its geometrical margins reach the side of the spouting vessel. Since the volumetric solids circulation through the spout/fountain system can be estimated to range to a few percents of the spouting gas flow rate, the fountain alone can pour very noticeable flows of material onto contiguous stages; then, its action should be limited by some mechanical device to restrain its hydrodynamic effect. Some devices, the so-called "Chinese hats" were presented in the sector literature (Mathur & Epstein, 1974) and tested in this unit. The experimental output was entirely disappointing, as these regulators failed in sufficiently defining the fountain shape, acted as a target for the particles propelled by the spout and interfered hydrodynamically with the gas flow in the freeboard, either when they were made of solid steel plates or wire mesh screen. To conclude, these devices are not advisable as internals in multistage spouted beds.

8.2.3 Freeboard baffles between stages

Each stage was segregated from its adjacent ones by side vertical baffles protruding down from the vessel top to a very short elevation over the bed free surface. The gap left had to assure solids flow only, depending on the continuous throughput rate; this gap between the submerged and the freeboard baffles has to be regulated to allow solids transfer by overflow, with restricted particle bounces from the fountain. Also bypass was minimized by this precaution. These flat and inexpensive devices were chosen to completely separate the freeboard into as many stage as the spouted bed design required. As a result, the action of each fountain (independently of its shape, thus gaining in spouting regime flexibility) was limited to its own stage. The use of these simple baffle repartition was observed to be fundamental for minimizing any interference between stages and enormously gain in stability.

8.2.4 Submerged baffles between stages

In principle, according to the fundamentals of fluidization, a multiple orifice spouted bed does not require a repartition between the annuluses. This consideration is also compatible with the particle vertical streamlines and the side-to-side homogeneous percolation of gas from a spout into the corresponding annulus. This assessment can be fully accepted when the system is operated batchwise and no net solids flow from one stage to the downstream one has to be steadily maintained. Practical reasons (easy start-up, spouting stability over time, independent gas flow rate regulation in each spouting module) have demonstrated that submerged baffles greatly help in defining the solids holdup in each stage. The separation of contiguous annular regions contributes in properly distributing the gas rate and giving origin to fully independent spouts. Conversely, if the holdup of solids is out of control, all the system stability may be affected.

8.2.5 Overall layout of a multiple spouting unit

In the recent past, in the frame of industrialization of a novel patented process for polyethylene terephthalate solid state polymerization (Cavaglià, 2003), a unit was conceived as a series of n-fluid beds (where n>5), operated either in turbulent fluidization or in spouting regime, where polyester beads with low intrinsic viscosity are heated-up and solid reacted in one equipment. A sextuple spouting demonstration unit was then built and operated as a prototype equipment for PET chips upgrading (Beltramo et al., 2009). The six modules were placed at identical elevation and positioned according to a 2x3 layout; the solids moved following a chicane path without being hindered by any internal repartition. The multiple spouted bed appeared advantageous in term of heat transfer efficiency (higher gas temperature at the inlet, thanks to a very short contact time between gas and heat sensitive solids) and generated good property polymers. However, the overall operation was troublesome because of difficult control on solids holdup and gas flow rate regulation. From that study sound design hypotheses were drawn to construct the experimental rig appearing in Figure 11A, whose main difference with respect to the previous industrial equipment consists in the possibility of positioning each stage at the desired elevation to facilitate the solids overflow to the downstream stage, as represented in the schematic of Figure 12. The final version of this experimental rig had the possibility of testing all the internals described above.

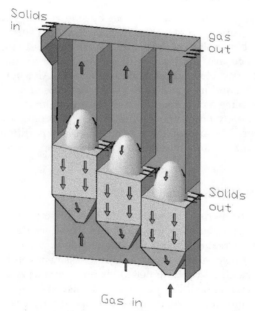

Fig. 12. 3D diagram of a continuously operating unit with a sloped layout of stages

The difference of level between units suitable for continuous and stable operations was evaluated both by running specific tests and by comparing these results against a simple correlation originated by estimating the angle of internal friction of the solids used in the experiments. The tests performed in the triple spouted bed unit aimed at devising the effect of increasing solids feed rates on bulk mass transfer from stage to stage, by measuring the angle of the bed surface with respect to the horizontal level, as well as the effectiveness of the internal baffle positioning. Figure 13 shows three different operating conditions at a mean (snapshot A), corresponding to a mean particle total residence time $\tau = 13.5$ min), high (snapshot B), τ = 4.5 min) solid rates and at a solids throughput far exceeding the nominal system capacity, as required by any foreseen process (snapshot C), τ = 2.5 min). The bed free surface slant increased to a maximum slope (about 15°, as a mean value between inlet and overflow sides) by increasing the flow rate. This angle is in the range of 1/2 to 1/3 of the solids repose angle, which can be measured following literature recommendation (Metcalf, 1965-66). By further enhancing the feed rate, the downcomer flooded unless further raised. The hydrodynamic slope that builds up at the bed surface caused the first stage to work with a solids depth quite higher with respect the last one, this difference increasing with the number of stages. It follows that the fluid dynamic control was much trickier and the overall spouting stability impaired.

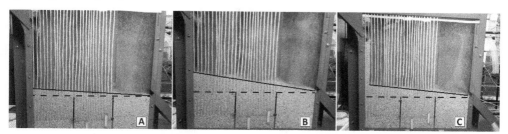

Fig. 13. Continuous operation in the three-module 0.20 m side spouted bed: – – – ideal solids free surface; ⎯⎯ actual solids surface at various solids flow rates: a) 2 kg/min, b) 6 kg/min and c) 10 kg/min, as given in the scheme of Fig. 11 B

This geometrical limitation was overcome by setting each bed at a minimum difference of level equal to:

$$\Delta H = D_c \tan \alpha \qquad (3)$$

where α is the angle formed by the actual solids surface with the horizontal level.

A rule of thumb suggests to determine the progressive vertical distance between adjacent spouted beds at:

$$\Delta H = 0.5\ D_c \qquad (4)$$

which, compared to the output of Eq. (3), leaves a safe operating margin.

As a conclusion, stable operations in a multiple square-based spouted beds require three types of flat internal baffles: one for properly addressing the solids feed to the bed surface, intermediate baffles in the freeboard to confine each fountain action, submerged baffles, each of them setting the solids overflow level from the upstream to the downstream stage. A non-interfering condition between stages was provided by generating a sloped cascade of independent spouting units.

8.3 Half-sectional spouted bed tests

A rig corresponding to half of the 0.20x0.20 m² spouted bed was built to represent a vertical section of the full unit. The axial sectioning included the orifice, the pyramid frustum and the constant cross section sector, thus originating a 0.20x0.10 m² column. A flat Perspex wall allowed a direct internal vision of spout, annulus and fountain, as already given in Figure 5.

Specific tests were carried out to demonstrate the close correspondence between data (H_m and U_{ms}) obtained in full and half-sectional columns. Approaching the maximum bed depth an underdeveloped and stable fountain was obtained with $U = 1.01\ U_{ms}$ (Figure 14 a)). Other runs highlighted a relevant bed expansion surmounting a submerged spout when the bed slightly exceeded the maximum bed depth (Figure 14 b)). Identical conditions also revealed occasional instability identified by a submerged wandering spout and some upper slugging, see Figure 14 c).

Fig. 14. Half sectional column tests at H: a) external spout and fountain formation at $H \approx 0.95\ H_m$, b) submerged spouting at $H \approx 1.05\ H_m$ c) spouting instability with spout wandering at $H \approx 1.05\ H_m$

Half-sectional units are suitable to measure particle cycle time, define solids streamlines, as well as visualize, at proper frame frequency, zones characterized by a high mixing degree. As far as the downward particle velocities are concerned, the considerations presented in the literature were taken into account, though obtained in semicylindrical vessels (Rovero et al., 1985). Figure 15 shows a sequence of snapshots which make visible the progressive motion of a tracer layer deposited on a fixed bed before starting the spouting process (first image). The second snapshot indicates that the tracer particles have maintained their position ahead of an external spout be formed. The third and fourth images indicate that particles move in the fountain in a piston flow fashion: then, local trajectories, their envelope (i.e. streamlines) and individual particle velocities can be defined. A minor portion of tracer only has been captured by the spout in the travel along the constant cross section of the

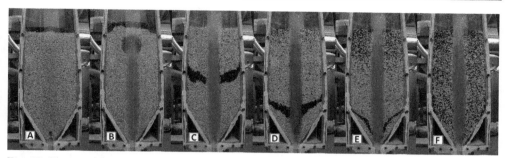

Fig. 15. The snapshot sequence indicates the motion of a tracer layer: A at t = 0 s; B at t = 1 s; C at t = 8 s ; D at t = 16 s; E at t = 22 s and F at t = 28 s

annulus. The fifth image gives evidence of noticeable shear acting on the particles, though no internal mixing occurs as far as particles enter the spout; then, quite an amount of particles is thoroughly mixed in the spout and fountain as they begin recirculating. The last image indicates an overall good mixing condition.

Spout diameter and spout profile were measured at the flat transparent wall. The experimental mean values were compared against the predictions given in the literature (McNab, 1972), with a difference of about 10%, which indicates that squared-based spouted bed findings do match the ones obtained in cylindrical vessels.

9. Fluid dynamics of solids in multiple spouted beds and its modelling

One of the main goals of a spouted bed cascade is to control the mixing degree of the overall system and possibly generate a piston flow of the solid phase to guarantee identical residence time to all particles. These evaluations are carried out by means of stimulus-response techniques after attaining steady state in a continuously operating unit. Since each individual spouted bed appears to have about 90% of its volume in a perfectly mixed state (Epstein & Grace, 2011), the recycle ratio (ratio of internal circulation in the spout referred to feed rate) leaves a small volumetric fraction of the annulus to operate as plug flow. Residence time distribution (RTD) studies in multiple spouted bed were presented in the literature (Saidutta & Murthy, 2000) in small rectangular columns having two or three spout cells. The absence of internals in this system brought to fountain wandering and excessive fountain heights that caused overall mixing higher than the one corresponding to the number of mixed units in series. A detailed RTD study on stable systems and the correlation of the experimental results with respect to the ones predicted by a model can give a relevant contribution in designing these units. Models has gained increasing importance by making use of direct measurements in the half-sectional unit, thus becoming fully predictive.

9.1 Residence time distribution function

The RTD curves represent an effective way to interpret the fluid dynamics of the solid phase in a multiphase continuously operating reactor. These functions describe the elapse of time spent by individual solids fractions in the system and can be modelled by a relatively simple combination of ideal systems, each of them describing a basic element (mixed or plug flow system, dead zone, bypass, recycling).

Two types of curves can be studied. The $E(t)$ function describes what a system releases instant after instant, i.e. the volumetric (or mass) fraction of particles whose residence time is between t and t+dt. The $F(t)$ function provides the integral of $E(t)$ ·dt and represents the fraction of elements whose residence time is lower than t.

The most direct way to trace an $E(t)$ curve makes use of a physical tracer, whose characteristics are identical or very close to the ones characterizing the bulk of solids travelling the system. Usually, two types of stimulus are adopted, a pulse (given by a definite amount of tracer introduced into the system in the shortest time) or a step (an abrupt change from the normal feedstock to an identical feed made of tracer only). The first one is generally the prompter to use. Right away after the introduction of the pulse, samples are taken at the system exit with a proper scrutiny degree and the tracer concentration is measured and recorded.

The two $E(t)$ and $F(t)$ functions are defined below:

$$E(t) = \frac{C(t) \cdot \dot{M}}{M_{tr} \cdot \rho_b} \tag{5}$$

$$F(t) = \int_0^t E(t)dt \tag{6}$$

with $C(t)$ being the concentration of solid tracer in the discharge, M_{tr} the mass of tracer injected, ρ_b the bulk density of tracer, $\dot{M}$ the mass flow rate of solids travelling the system.

A pulse function (called Dirac function, $\delta(t)$) is given by an instantaneous but finite entity (equal to unity) entering a system at t =0. The Laplace transforms of these functions allow the use of simple algebraic input/output relationships.

In our experiments, the pulse was obtained by quickly introducing in the feed a small amount of tracer (150 to 300 g, equivalent to about 1.5% of the total solids hold-up) made of PET chips doped with some ferromagnetic powder. After sampling the ferromagnetic PET chips were sorted out from the PET bulk by a magnet and their concentration calculated in each sample.

An example of RTD curves is given by Figure 16 which comparatively reports two $E(t)$ curves obtained at steady state from tests in the 3-module spouted bed operated with and without submerged baffles, according to the configuration appearing on Figure 11B. The two curves overlap almost perfectly to demonstrate that internal baffles do not alter at all the solids circulation in a multi-unit cascade. The same test also demonstrates that the external solids streamlines do not have any transversal (horizontal) component, as also visible from half column monitoring. The slope of solids at the free surface was modest in these runs, due to a very low throughput (2 kg/min). Thus, it is straightforward assessing that batch operations do not require any repartition between modules. From these considerations, the use of submerged baffles is beneficial to the start-up phase of continuous processes only and becomes fundamental as geometrical boundary to generate the configuration given in Figure 12.

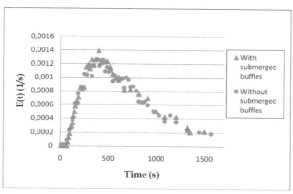

Fig. 16. Comparison between the RTD curves in the 3-module spouted bed with H/D_c=1.72 with and without submerged baffles

9.2 Modelling

The theoretical description of continuous units combines basic elemental models, whose combination gives origin to a system capable of generating an overall response to properly match the actual behaviour of the real system studied. A descriptive model can produce an output without having a strict link with the actual hydrodynamic behaviour and then has to make use of fitting parameters. This approach does not allow sound predictions or extension to more complex reactor structures. A much powerful tool is produced by conceiving a phenomenological model based on experimental observations. These models become fully predictive since all parameters are based on actual measurements. Then, a model interacts with scale-up procedures through the validity of the correlations used rather than its structure.

Dynamic responses were obtained by making use of a Matlab Simulink tool by generating model schemes as given in the below Figures 17 and 18. The fundamental modelling was based on a one-stage spouted bed; a multiple-cell system was then given by a cascade of basic units.

9.2.1 Descriptive model

The initial modelling started by considering an early dynamic description of spouted beds, where the overall behaviour can be portrayed by a well-mixed system with a minor (8 to 10%) portion of plug flow. The corresponding scheme given in Figure 17 A) includes the feed rate F_1, the bypass to fountain F_3 (which become negligible when the inlet diverting baffle is considered, as a consequence $F_2 \equiv F_1$ and $F_5 \equiv F_4$), the total circulation from spout F_4, the net discharge rate $F_7 \equiv F_1$ (for continuity) and $F_6 = F_4 - F_1$. From the experimental conditions F_1 is known and F_4 can be estimated by measuring mean particle velocity at the frontal wall of half-column. As far as the other parameters than appear in Figure 17 B) are concerned, t_d is estimated from particle circulation, $\tau_{annulus}$ follows from holdup in the annulus and $\tau_{spout+fountain}$ is calculated by difference from the known bed holdup. At usual H/D_c ratios adopted in cylindrical columns, the ratio of these two time constants approaches one magnitude order.

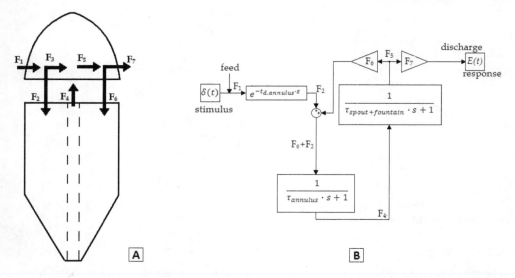

Fig. 17. Descriptive model of one stage two-zone continuously operating spouted bed: A): schematic of flow circulation between annulus and spout/fountain regions; B) Simulink model including pulse stimulus, delay at annulus entrance, two perfectly mixed regions for annulus and spout/fountain volumes and recirculation to bed surface.

Accepting this description also this model does not contain any fitting parameter, once the constitutive elements are assumed. Since the F_4 (internal circulation) to F_7 (net flow) ratio is very large, the overall system approaches a well-mixed unit and in this view the modelling is scarcely sensitive to the hydrodynamic description given to the annulus.

Figure 18 presents the comparison between experimental and modelling results for H/D_c = 1.72. The fitting is excellent, considering the time delay given by the minimum residence time of particles (t_d) and the well-mixed key dynamic component brought by spout recirculation.

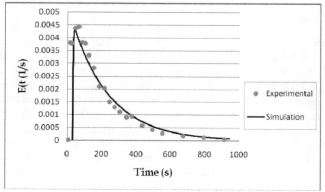

Fig. 18. Comparison between experimental data and the descriptive model

9.2.2 Phenomenological model

The relevant limitation contained by the above model consists in the fact that the annulus, representing the massive part of a spouted bed, has not been given a proper description. By observing it through the flat transparent wall of semicylindrical columns, particles show well-defined trajectories with scarce intermixing, according to consolidated findings (Mathur & Epstein, 1974).

The phenomenological description adopted in the updated model assumes that the squared-section of the annulus (with a cylindrical spout D_s) is divided into three axisymmetric zones, each of them having the same width according to the scheme given in Figure 19 A). Each of these regions receives from the fountain a solids flow rate proportional to its cross sectional area (F_{6A}, F_{6B} and F_{6C}, respectively moving from outside towards the spout). The flow fashion in each region is a piston with particle residence time $t_{d,A}$, $t_{d,B}$ and $t_{d,C}$, from the bed surface down to the cylinder-frustum junction, according to experimental observations. The mixing component acting in the annulus was concentrated in the frustum, which progressively discharges solids into the spout, depending on local streamline length. As a whole, this section was assimilated as far as its dynamics is concerned to a well-mixed volume, accounting to about 20% of the total holdup of the spouted bed. Also in this case, the effect of the F_4/F_7 ratio overcomes the sensitivity of other variables on the model, so that the ratio between frustum to parallelepiped volumes (i.e. plug to well-mixed volume ratio) is not relevant at all.

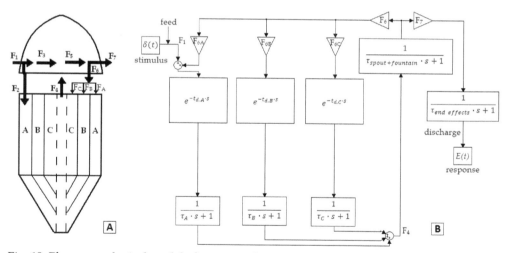

Fig. 19. Phenomenological model of one stage four-zone continuously operating spouted bed: A): schematic of flow circulation between annulus and spout/fountain regions; B) Simulink model including pulse stimulus, three parallel delay times in annulus, followed by three perfectly mixed regions in bottom frustum, one perfectly mixed zone in spout/fountain region and recirculation to bed surface. Small well-mixed volume accounts for sampling end effects.

Figure 19 B) presents the Matlab Simulink scheme, where the RTDF is generated by introducing the pulse into the sector A of the annulus (due to geometrical constrain of the inlet baffle). The overall solids flow rate from fountain travels three parallel annulus pistons, then each portion of solids enters the corresponding well-mixed portion of frustum, respectively characterized by a mean residence time estimated by observations at the flat frontal wall. The spout collects particles from the annulus and mix them in the fountain. A small well-mixed volume characterizes the solids sampling operation to account for end effects.

Figure 20 compares experimental results to the output of the phenomenological model. Any difference can be hardly noted with respect to the previous descriptive model output. Due to plug flow effect, a certain oscillation matching the cycle time frequency is observed. A short sampling time $\tau_{end\,effects}$ was sufficient to damp the greatest part of oscillation.

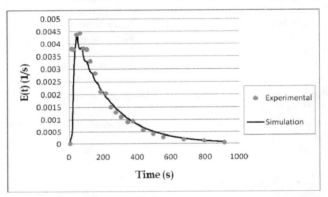

Fig. 20. Comparison between experimental data and the phenomenological model

9.2.3 Model validation for multiple units

The Matlab Simulink description conceived for the phenomenological model of one-stage spouted bed can be replicated a number of times corresponding to the number of cells included in a multistage system. An example is given in Figure 21 for a three-module

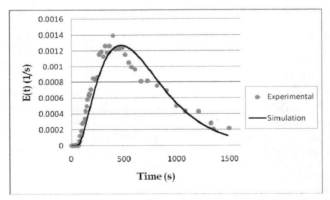

Fig. 21. Comparison between experimental and model RTDs in the 0.20 m side squared-based three-module spouted bed

spouted bed. The agreement is fully satisfactory, even though it may appear that experimental data anticipate the model output moderately and then a tail slightly higher than expected is displayed. This analysis could require to consider some direct bypass from fountain to downstream stage through the gap between submerged and freeboard vertical baffles and a small partially stagnant backwater, possibly existing along edges of frustum walls.

10. Final remarks on multiple bed start-up and shut-down

Both cold tests on the dual and triple module square-based spouted beds and industrial operations on the sextuple module demonstration unit (Beltramo et al., 2009) show that design criteria, operating conditions and step sequence must be defined carefully. In opposite case, start up and transitory from spouting onset in one module to overall stable spouting might be a serious issue. The key design parameters affecting start-up are:

- Cross-sectional area of the gap between vertical baffles: it should be designed to allow a maximum solids flow rate equal to the process solids nominal capacity with an excess of about 50% to avoid solids flooding (if too narrow) or bed emptying (if too large, because of direct discharge from fountain).
- Freeboard baffle height: it has to fully cover the fountain height at spouting onset and, if the structure of an industrial unit is considered, should reach the top deflectors of particles to avoid upper solids bypass.
- Cross-section of particle deflectors on gas phase path in each module (in the upper part of freeboard, above fountain projections): it has to be designed to set a progressive decrease of gas cross-sectional velocity and avoid any upward solids elutriation.
- The design of pyramid frustum base should consider a small gap between the orifice circumference and the slanted wall to help particle circulation. This gap is related to the average size of particulate material with a factor larger than 2. Several experiments have demonstrated that in case of local stagnation at the frustum bottom, this dead zone may spread up to the pyramid/parallelepiped junction. A careful bottom design and/or high gas flow rate counter this undesired phenomenon.
- Spouting modules have to be assembled at sufficient different elevation to guarantee a steady and even solids holdup. A rule of thumb drawn from experimentation suggests that $\Delta H \approx 0.5\,D_c$ surely gives origin to a non-interacting system and prevents bed emptying of downstream modules, or vice versa avoid extra hold up in upstream ones.
- Each module must be provided of independent gas flow regulation.

Given the fact that gas phase pressure drop is maximum at incipient start-up (up to 2 to 3 times of the gas pressure drop at stable spouting for conical or cylindrical shallow beds), to avoid oversize of circuit blower, with consequential dramatic increase of capital and operating costs, one can proceed according to two start up routes:

a. **Start-up with reduced solids holdup**: fill the first module with 1/3 to 1/2 depth with respect to the design bed load; start injecting gas till fountain formation, then begin filling the bed to reach the operating solids bed depth and then continue feeding solids at nominal flowrate to reach overflow discharge to the second module. Continue with the same procedure for all modules.

b.   **Start-up with full solids holdup:** fill each module with the design solids bed depth, provide the circuit with a capacitive booster section, suitable to inject high pressure gas to bed orifices for about 10 to 15 seconds and reach external spouting onset; then operate spouting process with the master circuit blower.

In a multiple-bed spouting unit, to have the whole series of spouted beds started up, one has to proceed with a start-up sequence, one by one, from the first bed to the last one, following the solids flow direction. Monitoring the gas phase pressure drop vs. time represent the trigger element for the control system to determine when the start-up of one module is completed and the situation is ready to move and start up next module. The b) procedure is safer since enough head is available to re-start the system in case of failure.

The above defined key design parameters are also suited for shutdown phase. As far as this procedure is concerned, one has to stop solids feeding, allowing holdup of each module to be processed at the steady state operating conditions (as to minimize off-spec), while decreasing each bed depth to a level lower than the overflow weir. At that point, side submerged baffles must be risen, while gas flow continues to be injected to each bed, so to have prompt residual solid emptying.

11. Conclusions

Spouted beds, throughout over half a century studies, have demonstrated to display very interesting features against bubbling conventional fluidization. Thanks their peculiar hydrodynamic structure a relevant gas rate can be saved also operating at the maximum spoutable bed depth. Again, the total frictional pressure drop across a spouting unit can be as lower as one-third of the one in a corresponding particulate material fluidization. The application of spouted bed to relatively coarse solids overcomes undesirable features characteristic of fluidization, namely random gas channelling and solids circulation, slugging and poor contacting between phases.

The scale-up issue of spouting units has remained open since their initial invention and the debate on whether prefer larger or multiple units has struck the opinion of scientists and technologists every time that this problem required a sound solution. This chapter has tackled the scale-up problem by opting for a square-based spouted bed geometry, since constructing a cascade of these vessels is economically advantageous and much more effective for the solids fluid dynamics as well as for insulation problems.

Following the experience gained during an industrial demonstration project, encouraging results came by evaluating the product quality; the potential performance of a multiple spouted bed was thus confirmed. Nevertheless, this unit required an excessive attention to govern its stability over time and the need of several improvements was highlighted.

A new research project required the construction of several apparatuses to pursue a comprehensive strategy which has aimed at:

- comparing the fundamental operating parameters of square-based spouted beds with the corresponding values characteristic of conventional cylindrical columns,
- carrying out an adequate experimental scale-up,

- demonstrating the achievement of fully stable operations by introducing novel concepts (stage segregation with internal baffles, sloped cascade of stages) in designing a multi-cell equipment,
- obtaining a plug flow of solids with a sufficient number of stages, which may implement process scale-up at the same time,
- modelling single unit and multiple square-based spouted beds to predict solids hydrodynamics.

The final structure of a cold model apparatus has demonstrated the achievement of all the listed goals.

12. Acknowledgments

The economical support granted by ENGICO Srl (LT– Italy) allowed construction of the equipment and the research fellowship to one of the authors (M. C.). The authors wish to tank Mr. Alfio Traversino for patiently and carefully constructing the three-module experimental rig.

13. Nomenclature

Ar	Archimedes number
C	concentration of solid tracer in the discharge, kg / m^3
D_c	column diameter, m
d_i	inlet diameter, m
d_p	particle diameter, m
D_s	mean spout diameter, m
$E(t)$	E function
$F(t)$	F function
H	bed depth, m
H_m	maximum spoutable bed depth, m
ΔH	gap between adjacent units, m
$\dot{M}$	mass flow rate of solids, kg / s
M_{tr}	mass of the tracer, kg
ΔP_M	maximum pressure drop across bed, Pa
ΔP_S	spouting pressure drop across bed, Pa
U	fluid velocity, m / s
U_{ms}	minimum spouting velocity, m / s
U_{onset}	fluid onset velocity, m / s

Greek letters:

α	free surface slope, deg
$\delta(t)$	Dirac function
λ	shape factor
ρ_f	fluid density, kg / m^3
ρ_b	bulk solids density , kg / m^3
ρ_s	actual material density, kg / m^3
τ	mean residence time, s

14. References

Beltramo, C.; Rovero, G. & Cavaglià, G. (2009). Hydrodynamics and thermal experimentation on square-based spouted beds for polymer upgrading and unit scale-up. *The Can. J. Chem. Eng.*, Vol.87, 394-402

Cavaglià, G. (2003), "Reactor and process for solid state continuous polymerisation of poly-ethylene terephthalate (PET)" Patent EP 1576028 B1

Chatterjee, A. (1970). Spout-fluid bed technique. *Ind. Eng. Chem. Process Des. Develop*, Vol.9, 340-341

Epstein, N. & Grace, J. (2011). *Spouted and spout-fluid beds*. Cambridge Univ. Press, ISBN 978-0-521-51797-3,New York

Gishler, P.E. & Mathur, K.B. (1957). Method of contacting solid particles with fluids. U.S. Patent No. 2,786,280 to National Research Council of Canada

Grbavčić, Ž.B.; Vuković, D.V.; Hadžismajlovic, D. E.; Garić, R. V. & Littman, H. (1982). Fluid mechanical behaviour of a spouted bed with draft tube and external annular flow. *2nd Int. Symp. on Spouted Beds, 32nd Can. Chem. Eng. Conf.*, Vancouver, Canada

Geldart, D. (1973). Type of gas fluidization. *Powder Tech.*, Vol.7, 285-292

Malek, M.A. & Lu, B.C.Y. (1965). *I&EC Process. Des. Develop.*, Vol.4, 123-127

Mathur, K.B. & Epstein, N. (1974). *Spouted beds*, Academic Press, ISBN 0-12-480050-5, New York

McNab. G.S. (1972). Prediction of spout diameter. *Brit. Chem. Eng & Proc. Techn.*, Vol.17, 532

McNab, G.S. & Bridgwater, J. (1977). Spouted beds – estimation of spouting pressure drop and the particle size for deepest bed. *Proc. of European Council on Particle Technology*, Nuremberg, Germany

Metcalf, J. R. (1965-66). The mechanics of the screw feeder. *Proc. Inst Mech. Eng.*, Vol.180, 131-146

Murthy, D.V.R. & Singh, P.N. (1994). Minimum spouting velocity in multiple spouted beds. *The Can. J. Chem. Eng.*, Vol.72, 235-239

Piccinini, N. (1980). Particle segregation in continuously operating spouted beds, In: *Fluidization III*, J.R Grace & J.M. Matsen,(Eds), 279-285, Plenum Press, ISBN 0-306-40458-3, New York, USA

Rovero, G.; Piccinini, N. & Lupo, A. (1985). Vitesses des particules dans les lits à jet tridimensional et semi-cylindriques. *Entropie*, Vol.124, 43-49

Saidutta, M.B. & Murthy, D.V.R. (2000). Mixing behavior of solids in multiple spouted beds. *The Can. J. Chem. Eng.*, Vol.78, 382-385

4

Optimal Synthesis of Multi-Effect Evaporation Systems of Solutions with a High Boiling Point Rise

Jaime Alfonzo Irahola
Universidad Nacional de Jujuy
Argentina

1. Introduction

In the past, optimization had been studied only for typical flowpatterns like forward and backward feed. To make the choice between them, simple rules were applied based on the viscosity and the temperature of the initial dilute solution (T_F). Thus, forward feed was usually favored for the evaporation of low-viscous hot solutions featuring a temperature $T_F>T_P$, where T_P is the desired temperature of the final product. By doing so, the liquid heating load is largely cut down. In turn, backward feed was recommended for heavy-viscous cold liquors.

Moreover, a few contributions to the optimal synthesis of multiple effect evaporator systems (MEES) have so far been published. Most of the previous papers was focused on the analysis rather than the synthesis of evaporation systems. They generally assumed that important structural variables like vapor and liquid flowpatterns and the number of effects are all known data though they drastically change the performance of a MEES.

Nishitani and Kunugita (1979) first presented a multiobjective problem formulation to determine the optimal flowpattern of a multiple-effect evaporator system. However, they did not consider stream mixing/splitting. In addition, the solution method performed one-by-one the simulation of the MEES for all possible flowpatterns. More recently, Hillebrand and Westerberg (1988) developed a simple model to explicitly compute the utility consumption for multiple-effect evaporator systems exchanging sensible heat with outside streams. In turn, Westerberg and Hillebrand (1988) introduced the concept of "heat shunt" to derive the best liquid flowpattern in a heuristic way. Nonetheless, major assumptions like constant boiling point elevation, no liquid bypassing and negligible heat of mixing somewhat limit the applicability of their findings.

To resolve the problem presented here has been used a mathematical model rigorous previously developed (Irahola & Cerdá, 1994). It considers the possibility of optimizing the variables that you want to. This has allowed that the model is used to solve various types of problems, namely: simulation, optimization, optimal synthesis and optimum partial reengineering restrictive of multi effect evaporation system (MEES) (Irahola, 2008). The mathematical model developed is the MINLP and solved using commercial software. The approach was successfully applied to three industrial problems. Depending on the feed and the product temperature, the optimal configuration uses a distinct liquid flowpattern that often differs from the conventional forward and backward feed and leads to reasonable savings (Irahola & Cerdá, 1996)

Among the results should be noted that: the splitting of the flow of live steam can be a better alternative than the traditional cascade of steam; the best fixed cost curve is not always a monotonous increasing; the correct distribution of the areas of the effects of the MEES, the appropriate operating conditions and the correct choice of liquid and vapor flowpatterns, are the determining factors in the optimal design of the MEES.

Perhaps the greatest disadvantage found using the method proposed here to solve the formulated mathematical model (MINLP type) for optimal synthesis of the MEES is the presence of numerous local optimal what makes it difficult to obtain optimal Global.

2. Evaporation of an aqueous solution of caustic soda

Among the solutions of industrial interest that present a high increase in boiling point are sodium hydroxide solutions (caustic soda). The concentration of these substances by evaporation, presents significant disadvantages due to the characteristics of the caustic solutions, namely:

• Have a high boiling point elevation (BPE) which implies a great loss in the temperature difference available.

• Concentrated solutions are highly viscous, which severely reduces the rate of heat transfer in natural circulation evaporators.

• They can have detrimental effects on steel, causing what is called caustic fragility. In addition, they may require removal of large amounts of salt when the solution is concentrated.

Since the transfer of heat (U) of liquor film coefficient, depending on the speed of the caustic solution through tubes (among other variables), usually, seeks a high speed in order to obtain a large coefficient (Kern, 1999).

According to the literature, it has taken as standard for the concentration of caustic soda, a evaporation system of two or three effects operate in backward feeed (Kern, 1999). In this study, found that the structure in counterflow or backward feed, obviously presents a high performance, but is not the best. In order to confirm what was said, is going to solve a problem.

3. Optimal synthesis of a multi-effect evaporation system for the concentration of caustic soda

3.1 Problem

Find the optimal MEES to concentrate 30040 lb/h (13626 kg/h) of an aqueous solution of sodium hydroxide from 10 to 50% by weight. The type of used evaporator is long vertical tube with forced circulation. Available in the plant: live steam boiler to 63.69 Psia (4.48 kg/cm2). The allowable minimum absolute pressure in an effect is of 1.942 Psia (0.1365 kg/cm2) (Geankoplis, 1983).

3.2 Mathematical model

The scope of the rigorous mathematical model is limited by the following assumptions adopted in the formulation:

a. In each effect, the vapor and liquid phases are in equilibrium.
b. A solid phase never arises in any effect.
c. The impact of the hydrostatic head on the liquid boiling point is neglected.
d. There is no leakage or entrainment.
e. Heat losses from any effect need not be considered.
f. The steam always condenses completely.
g. Subcooling of the condensate is very small.
h. Flow of noncondensables is negligible.
i. The concentrated final product is withdrawn from a single effect which in turn does not transfer liquid to any other one.
j. If necessary, you can use a heat exchanger or condenser so that the product go out to the preset temperature (Tp).
k. Not consider any type of pump between the effects.

To solve the mathematical model and find the optimal design should be available before, the mathematical expressions for all dependent variables: enthalpy of steam (H), enthalpy of solution of soda caustic (h), latent heat of vaporization of the water (λ), overall heat transfer coefficient (U), temperature of the solution in the evaporator (T) and costs of forced circulation evaporator, barometric condenser multijet, surface condenser and heat exchanger. In general, useful information is available in graphics, which presented various authors cited in the bibliography, but there are no equations of those curves. These situations and other problems are resolved below. According to Standiford (1963), in forced circulation evaporators, film transfer coefficient (h) on the side of the liquid, can be calculated from the conventional Dittus-Boelter equation for forced circulation when there is no boiling.

$$\frac{hD}{k} = 0.0023 \left(\frac{DG}{\mu}\right)^{0.8} \left(\mu \frac{Cp}{k}\right)^{0.4} \tag{1}$$

If there is this equation for two points and combine both equations, you can find the functionality of h_1 with respect to another point (2) as:

$$h_1 = h_2 \left(\frac{\mu_2}{\mu_1}\right)^{0.4} \left(\frac{C_{p1}}{C_{p2}}\right)^{0.4} \tag{2}$$

As the overall heat transfer coefficient U is practically determined by the film coefficient h fluid side, the above equation can be used to obtain a correlation for U. In Geankoplis for T=105 °F and X = 0.5, data is U = 400 (Btu/h ft^2 °F). T and X is obtained from a graph (Horvath, 1985) μ_2 = 22.84 centipoise. Then, for five values of concentration and six temperature values are obtained from graphics (Horvath, 1985) the values of C_{p1} and μ_1. With the data obtained can be calculated according to the above expression, the overall heat transfer coefficient U_1 with reference to U_2. Thus, U is plotted vs. X (Fig. 1) and U vs. T (Fig. 2). Finally, using the triple X, T, U_i, can be found by regression, the functionality of U in terms of concentration and temperature:

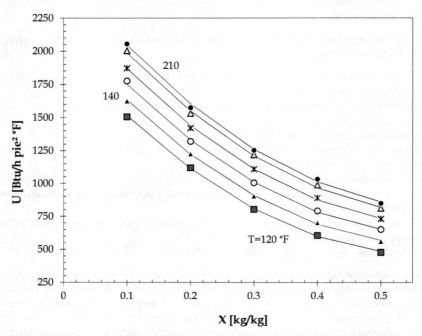

Fig. 1. Overall heat transfer coefficient correlation as a function of solute concentration.

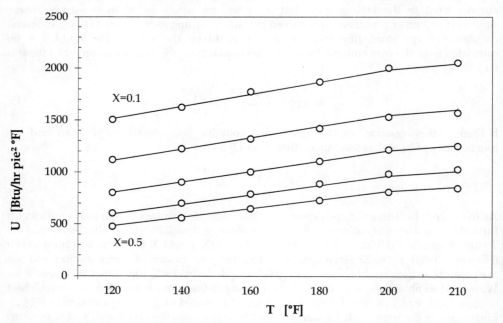

Fig. 2. Overall heat transfer coefficient correlation as a function of temperature.

$$U = 1254.780865 - 4954.6700\,X + 4832.059524\,X^2 + 6.321549\,T$$
$$- 4.30974\,XT \left[\frac{Btu}{h\,ft_2\,°F}\right] \tag{3}$$

where: X: mass fraction and T [=] ° F

The regression function is chosen to achieve maximum correspondence with the data, but at the same time trying to maintain, if possible, the simplicity. Nevertheless, we could not avoid the bilinear term. Figures 1 and 2, you can appreciate the good fit of the correlation found for U.

Also, to represent the enthalpy of solution, depending on the concentration and temperature, it became a non-linear regression with data from Foust (1980) (Fig. 3):

$$h = -10.250000 - 319.591837\,X + 939.795918\,X^2$$
$$+ 0.963929\,T - 0.335714\,XT \left[\frac{Btu}{lb}\right] \tag{4}$$

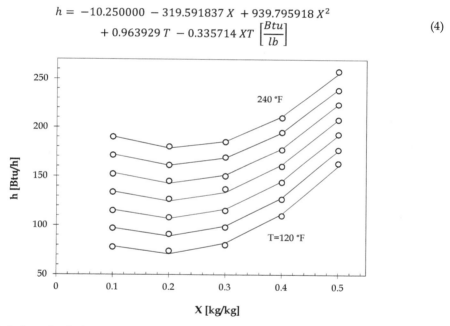

Fig. 3. Enthalpy of solution

The functional expression of the cost of barometric condenser multijet (C_{CBM}) is obtained by regression analysis of the chart presented by Peters (1991). Again, searching for mathematical expressions that minimize non-convexities of the mathematical model, we found that a quadratic expression represents excellent, the cost curve in the range of work desired (Fig. 4) (1992):

$$C_{CBM} = 4,753104 + 2,480885\,w + 0,281818\,w^2 \left(10^3\ USD\right) \tag{5}$$

Where: w [=] gpm. The flow rate of cooling water (w) required in the condenser is directly given by (Kern, 1999):

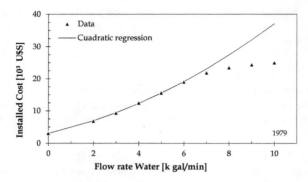

Fig. 4. Installed Cost of multijet barometric condenser.

$$w = \frac{Q}{500(T_s - T_w - t_a)} \text{ (Gpm)} \tag{6}$$

where:

Q = heat load, Btu / h
T_s = temperature of saturated steam, °-F
T_w = temperature of cooling water, °-F
t_a = 15 °-F = degree of approximation at T_s

The cost of forced circulation evaporator is obtained based on information reported by Maloney, 2008:

$$CE_{FC} = 2420.5\, A^{0.7121}\ U\$S \tag{7}$$

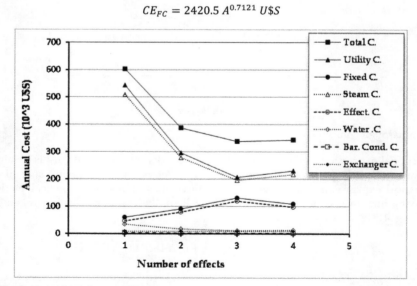

Fig. 5. Evolution of total operating costs.

The information of the data used in the resolution of the problem is presented in table 1. Also, the thermodynamic properties of the remaining functions and costs of services and equipment are presented.

In order to consider probable situations that could be presented in the industry, will study the cases in which the temperature of the weak solution (feed) is equal, higher or lower than the temperature of the strong solution (product). In the three cases, only change the values of the inlet of the weak solution temperature. The rest of the parametric conditions remain fixed.

Feed and product conditions				
Item	Feed		Product	
Flow rate lb/h (Kg/h)	30040	(13626)	60080	(2725.2)
Weight fraction	0.1		0.5	
Temperature °F (°C) — Case I	180	(82.2)	130	(54.4)
Temperature °F (°C) — Case II	130	(54.4)		
Temperature °F (°C) — Case III	80	(26.7)		
Operation conditions				
	Temperature °F (°C)			
Steam	296.6 (147.0)			
Cool water	(min) 89.6 (32.0)		(max) 107.6 (42.0)	
Effect	(min) 125.0 (51.7)		(max) 294.6 (145.9)	
	Minimum allowable temperature difference °F (°C)			
Condensers	18.0 (10)			
Heat Exchangers				
Thermodynamic properties				
Specific vapor enthalpy	$H_i = 1075 + 0.3466\,Tv_i$ [Btu/lb]			
Solvent Latent Heat of Vaporization	$\lambda_i = 1104 - 0.65\,Tv_i$ [Btu/lb]			
Operating temperature at effect i	$T_i = (1 + 0.1419526\,X_i)Tv_i - 9.419608\,X_i + 271.3627\,X_i^2$ [°F]			
Costs of utilities and equipment				
Steam	2.922×10^{-6} (USD/Btu) 2.104E-2			
Cool water (305 K)	1.952×10^{-7} (USD/Btu) 1.405E-3			
Surface condenser	$1092.83\,A^{0.65}$ (USD)			
Heat exchanger	$1144.16\,A^{0.65}$ (USD)			

Table 1. Data for example and, functional expression of thermodynamic properties and the costs of utilities and equipment.

4. Analysis and discussion

4.1 Case I. Feed temperature higher than the temperature of the product ($T_F > T_P$)

4.1.1 Comparative analysis of the optimal solution found

Adopted T_F = 180 °F (82.2 °C). Coinciding with the generally accepted criterion of optimality for the evaporation of caustic soda, has been found that the optimum number of effects of the MEES is equal to three (Fig. 5). However, a mixed structure {2,1,3} has been found in the path of the current liquid instead of backward feed. The feed stream enters to the effect 2 and then continuous countercurrent to the effect 1. Then go out and circulates in forward feed to the effect 3. (mixed liquid flowpattern). In this new structure that is presented (Fig. 6), we see significant increases in boiling point of the solution: 10.5 °C and 40 °C, in effects 1 and 3 respectively.

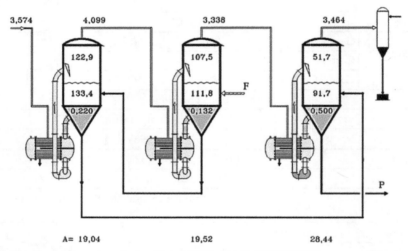

Fig. 6. Optimal configuration three-effect MEES. (TAC=337835 USD)

The optimal solution will be the one with the lowest total annual cost (TAC). From this point of view, the classical structure proposed as an optimum solution: evaporation system of three effects of equal area arranged backward feed (BFA), is 4.31% more expensive than the optimal solution (SO) (Fig. 7). More, even if it is allowed to in the structure backward feed, the effects have distinct areas (BF), do not get a better result that the optimal solution found. The difference in cost is 4.03%. In the figure 7, we also present results for a forward feed evaporation system. This structure, in its classic form forward feed with effects of equal area (FFA), is 7.91% more expensive. Which could corroborate, in some way, because in the past the BFA structure was preferred. If is allowed that the effects have different areas in the structure forward feed (FF) is very interesting the result obtained. The correct distribution of areas has led to a decrease in the total cost. The decrease is so great that, now, the FF structure is better than any of the above structures in backward feed. Its total annual cost (TAC) is only 2.95% greater than the of the optimum solution found (Fig. 7). In general, it appears that whatever the structure of MEES, the operating cost is significantly greater than the fixed cost (Fig. 8). It is approximately 65% of the total cost.

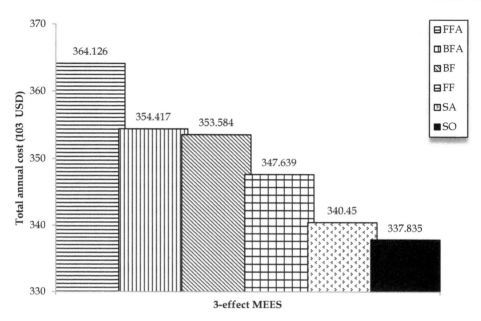

Fig. 7. Comparison with Typical Flowpatterns.

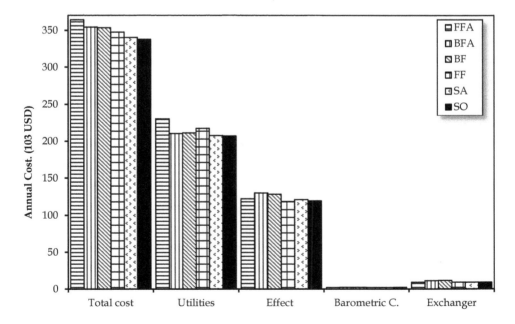

Fig. 8. Relative incidence of Operating and fixed Costs

4.1.2 Impact of the flow pattern

In this case, the trajectory of the liquid stream is the determining factor in the performance of a given MEES. The benefit achieved is even greater than obtained by allowing the effects having different areas with each other. That said, is based on the result of the structure SA. This has the same flow pattern that the optimal solution, but the effects of evaporation system are of equal area. The increase in cost is only 0.8% compared to MEES optimum (Fig. 7). From the practical point of view, the alternative SA may be the best option.

4.1.3 Profiles of the structural and parametric variables

As it will be seen later, only in this case it can be seen some regularity in the curves of the structural and parametric variables. Furthermore, after reaching the optimal point generally the next curve is anomalous with respect to the preceding ones. In the last effect occurs the maximum concentration jump (ΔX) of the solution (Fig. 9). At the same time, it has the maximum area as shown in the curves of the 1 to 4-effect optimal MEES (Fig. 10). The flow rate of steam produced in the effects is approximately the same. However, this does not apply to MEES with greater number of effects than the optimal. (Fig. 11). The profile of the temperature to the optimum MEES of a different number of effects, does not have a regular aspect. However, the maximum temperature jump occurs in the last effect (Fig. 12).

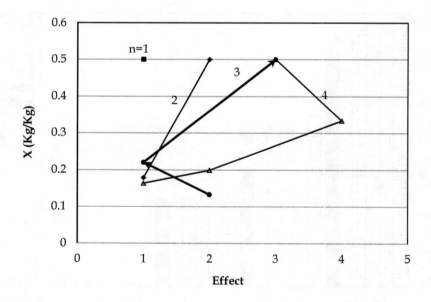

Fig. 9. Concentration profiles in the 1 to 4-effect MEES. (n: number of effects)

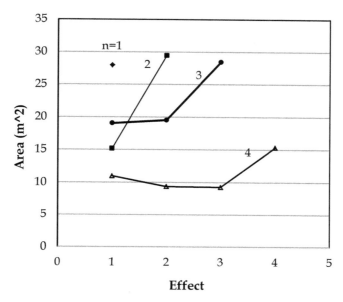

Fig. 10. Area profiles in the 1 to 4-effect MEES (n: number of effects).

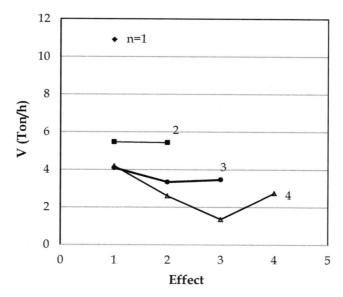

Fig. 11. Flow rate profiles in the 1 to 4-effect MEES (n: number of effects).

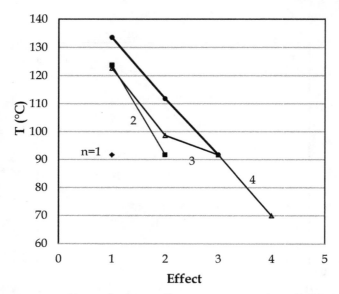

Fig. 12. Temperature profiles in the 1 to 4-effect MEES (n: number of effects).

4.2 Case II. Feed temperature equal than the temperature of the product ($T_F = T_P$)

4.2.1 Impact of the splitting of the live steam flow rate on the optimal solution

The result obtained when TF = TP, is different, not only to the found for the case I, but also with respect to the classical position. Found structure is highly innovative and simple in its conception.

If you look at the evolution of the total annual cost (TAC) curve of the 1, 2, 3, 4 and 5 effect Optimal MEES, it was found that the four-effect MEES is that of lower cost (optimal quasi-global) (Fig. 13). The flow pattern is backward feed and in this aspect, this result coincides with the classical motion, but not with the number of effects: proposed here a four-effect MEES, instead of three. However, this new proposal, would not be really the best alternative, if it was not associated to the new steam flow pattern proposed (Fig. 14). It emphasizes, splitting in the live steam flow pattern, it enters parallel to the effects 1 and 2; the by-passing effect 2 by the vapor stream from effect 1 and finally, the mixing of vapor streams from effects 1 and 2 for heating effect 3.

Against, this new trajectory of the flow of steam, first doubt that occurs, is the performance of this configuration against the unifilar cascade of high thermodynamic efficiency.

4.2.2 Comparative study of proposed flow pattern with respect to traditional configurations

For the purpose of explaining the improvement achieved, compares the structure backward feed with effects of different areas (BF) and the optimal solution (SO). Both structures have equal number of effects and same trajectory of the liquid flow and only differ in the

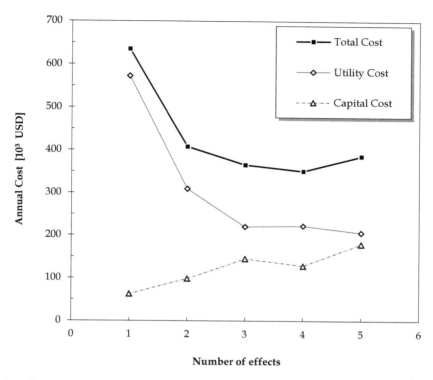

Fig. 13. Relative incidence of Operating and fixed Costs.

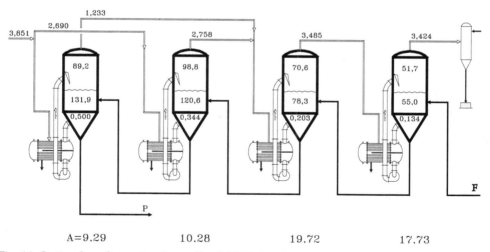

Fig. 14. Optimal configuration four-effect MEES. (TAC=351082 USD)

trajectory of the flow of steam. Two comparisons were made: one relating to the cost of auxiliary services and the other with respect to the cost of the effects, which is almost all of the fixed cost.

4.2.3 Energy efficiency and fixed cost of the traditional structure

The BF structure has a higher efficiency since the cost of auxiliary services is 174963 USD, 21.30% less than the cost for the SO. On the other hand, the cost of the effects is 193008 USD, i.e. 167,8% of the respective cost observed in SO. The net result of the comparison of the total costs, indicates that the BF structure is 8.9% more expensive than SO. With these results, following the classical position we can say, that BF 4 effects is not better than the optimum found (SO), because the MEES should be a structure BFA 3 effects, not four, which was used to compare. Therefore, will be then verified the validity of this rule, for the case study.

4.2.4 Difference in the number of effects due to non-traditional flow patterns

The optimal number of effects found by the mathematical model does not coincide with the optimal number for BFA and FFA traditional structures or even structures BF and FF. This explains why in the absence of a mathematical model to explore the multiple alternatives of design, the best answer to the problem was until now, a countercurrent system.

It was found that the optimum number of effects to structures backward feed and forward feed is 3. However, the developed model proposes the structure SO of 4 effects as the best solution. Therefore, to verify the quality of it, is advisable to compare the best results found for each structure.

4.2.5 The optimal solution compared to traditional structures

The total cost of the BFA MEES is 4.2% more than the optimal. Therefore, the structure and number of effects, traditionally proposed do not seem to be the most appropriate. Then, one might think that if you remove the restriction of equal area of the effects, could be improved, significantly, the current result. The results show that the BF structure of three effects is 4.1% more expensive than SO (Fig. 15). However, despite the difference in the number of effects, is convenient to analyze in more detail these recent results.

Structurally, BF and SO differ only in the flow pattern of steam. The cost of the auxiliary services of BF is 1.0% lower than the SO. On the other hand, the fixed cost is 13.4% greater determining to SO submit one minor TAC (Fig. 16).

4.2.6 Profiles of the structural and parametric variables

This case is characterized because the profiles of the process variables, for the various intermediate optimal MEES, they have no similarity among themselves. In particular, notes that the optimal solution presents the most discordant curve with respect to the others.

The temperature profile is irregular with temperature differences between effects non-uniform, being the most important jump located between 2 and 3 effect, following the drop of temperature effects 3 and 4, both heated with secondary steam. (Fig. 17). The greater temperature difference between the heating steam and the solution to evaporate, occurs in

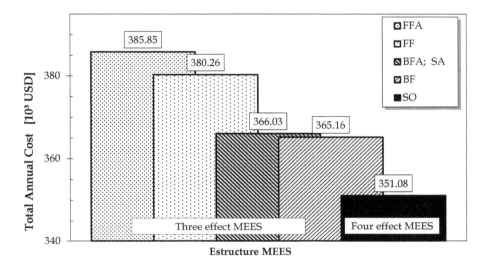

Fig. 15. Comparison with Typical Flowpatterns.

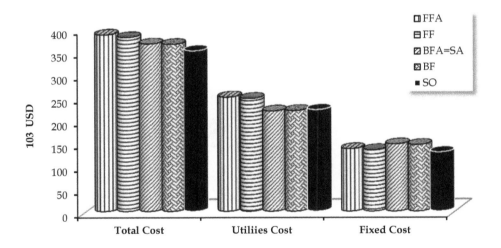

Fig. 16. Comparison with Typical Flowpatterns (Case II).

the effect 2 of the four effect MEES (Fig. 18), through the use of live steam in the effect. Thermal jumps that are achieved with the optimum structure are higher that in the triple and quintuple-effect MEES. In addition, it should be noted that the thermal jump in effect 2 is almost doubled with respect to other effects.

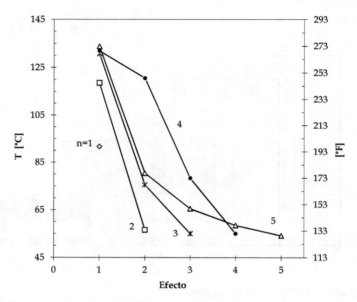

Fig. 17. Temperature profile optimal MEES for n=1 to 5 effects.

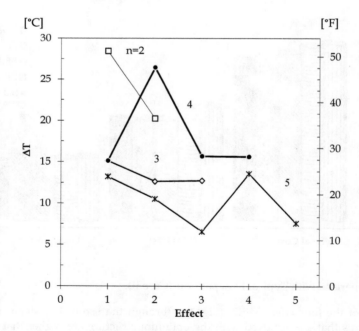

Fig. 18. Temperature difference profiles between the condensation and evaporation chambers in each effect for 2, 3, 4 and 5 effects.

Similarly, the profile of the area for each MEES, is far from being uniform, with the highest values located in the lower thermal effects. (Fig. 19). Should be mentioned, that the problem had been resolved for a fixed range area (100, 1000 ft²).

The flow rate of solvent evaporated in each effect is approximately the same in the double and triple-effect MEES (Fig. 20), where the chosen structure is backward feed. But in the optimal solution and after this, the values of vapor flow rate are far from each other. Although it can be seen that, in SO the curve is regular and decreasing with temperature

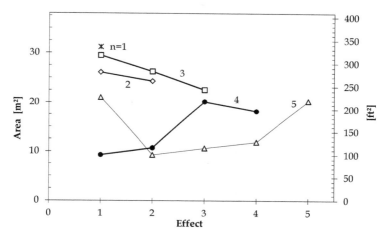

Fig. 19. Area profiles in the 1 to 5 effect MEES (n: number of effects)

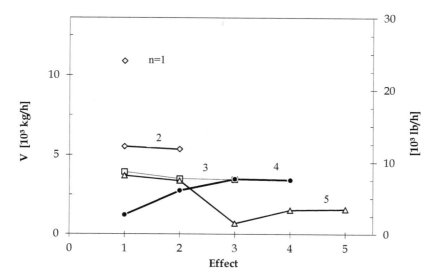

Fig. 20. Flow rate profiles in the 1 to 5-effect MEES (n: number of effects).

effects, i.e. the greater evaporated flow rate occurs in the effect 4. The profile of the global coefficient U is similar: lower in effect 1 and higher in the effect 4. It is important to clarify that, the values of the flow rate of steam produced in each effect are not directly indicative parameters. Yes it is a relative measure, for example, the percentage of the current liquid evaporates. Thus, the percentage amount of solvent evaporated, with respect to the flow rate of solvent that enters each effect, is maximum in the second effect (41%), followed in decreasing order, the third effect (34%), the first (31%) and finally the fourth (25%).

The concentration curve of the solution is monotone increasing, considering the path of the liquid stream (Fig. 21). The biggest jump in percentage concentration occurs in the second effect (69 %), then in descending order, the third effect (51 %), the first (45 %) and the fourth (34 %).

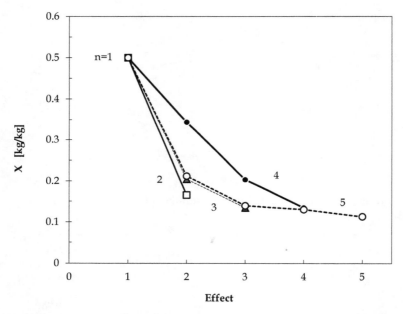

Fig. 21. Concentration profiles in the 1 to 5-effect MEES

The recent analysis of incremental concentration, the carried out for the steam produced in the effects, and the biggest jump thermal observed in effect 2 of the Optimal MEES show the importance of this effect. Its presence is the root cause of the improvement achieved in SO. This is achieved, thanks to the optimal design of the MEES, which allows an adequate relationship of the variables that define the system.

4.3 Case III. Temperature of the weak solution lower than strong solution ($T_F < T_P$)

4.3.1 Uniqueness of the cost curves

This case presents great similarity with the Case II. The optimal effect is the same (4) and intermediate structures found for $n \leq 4$ are almost identical. The best configuration for the liquid stream, given by {4,3,2,1}, as in Case II, use live steam to heat the first two effects.

Evolution of cost curves show a singularity with respect to the previously analyzed cases. Now, to increase the number of effects to reduce the TAC, the consumption of live steam begins to increase (Fig. 22) rather than continue to decline with the addition of a new effect, and the cost of the effects decreasing, rather than continue to increase. Isn't traditional behavior: a curve monotone decreasing for the cost of the live steam and one monotone increasing for the cost of the effects. Here is not complied with this scheme because of the significant reduction of the U coefficient and the high increase of the boiling point (BPE) in the effects of greater concentration. It is now "pays" with live steam part of the savings in the areas of effects.

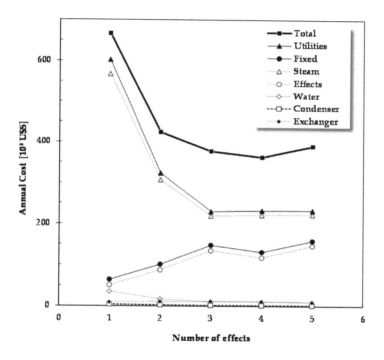

Fig. 22. Evolution of the TAC and its component terms with the number of effects (Case III).

4.3.2 Structure and distribution of temperature, concentration and heat transfer area

A structural analysis allows us to appreciate that the cuadruple-effect MEES results have added an effect, between the first and the second of the triple-effect MEES (Figs. 23 and 24, Table 2.).

Looking to reduce the total cost, the new effect requires one of the lowest values of area of heat transfer (Fig. 25). For this purpose, it operates with a large temperature difference; almost double that for the remaining effects (Fig. 26). On the other hand, contrary to expectations, their presence causes a slight increase in the consumption of live steam and allows at the same time, an increase in the thermal jump in the last two effects of, approximately, 6 °F (3.3 °C).

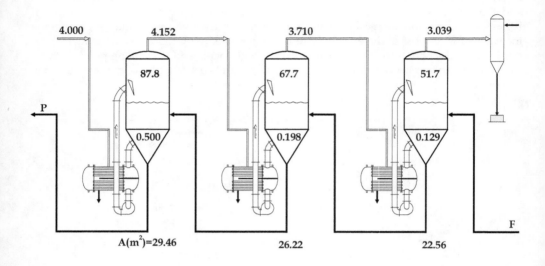

Fig. 23. Optimal Structure for three effects. (Local optimun).

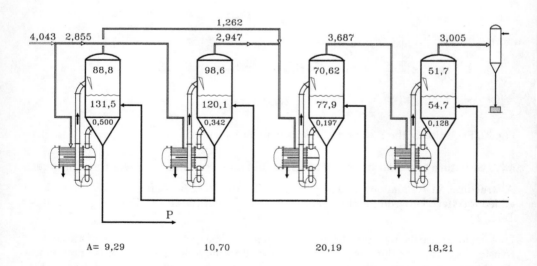

Fig. 24. Optimal Structure for four effects. (Optimal Solution).

Alternative Optimal Solutions (Case III)							
Equal area effects MEES: SA							
Effect	T	Tv	Vs	L	V	X	F
(i)	[°C]			[10³ kg/h]		[kg/kg]	[m²]
1	137.99	94.92	4.105	0	3.232	0.500	
2	128.93	107.28	0	0	2.979	0.340	15.33
3	80.62	73.39	0	0	3.665	0.195	
4	54.72	51.67	0	13.626	2.971	0.128	
Backward feed MEES. Optimal area effects: BF							
Effect	T	Tv	Vs	L	V	X	F
1	130.40	87.83	4.000	0	3.232	0.500	29.47
2	74.99	67.71	0	0	3.710	0.198	26.22
3	54.76	51.67	0	13.626	3.039	0.129	22.57
Backward feed MEES. Equal area effects: BFA							
Effect	T	Tv	Vs	L	V	X	F
1	128.39	85.95	3.984	0	4.126	0.500	
2	73.12	65.84	0	0	3.719	0.199	26.54
3	54.77	51.67	0	13.626	3.055	0.129	
Forward feed MEES. Optimal area effects: FF							
Effect	T	Tv	Vs	L	V	X	F
1	132.77	127.95	4.822	13.626	3.517	0.135	20.37
2	116.53	107.02	0	0	3.714	0.213	18.96
3	91.67	51.67	0	0	3.671	0.500	32.69
Forward feed MEES. Equal area effects: FFA							
Effect	T	Tv	Vs	L	V	X	F
1	135.12	130.25	4.870	13.626	3.530	0.135	
2	121.62	111.99	0	0	3.691	0.213	24.40
3	91.67	51.67	0	0	3.680	0.500	

Table 2. Results of alternative evaporation systems. (Vs: Live steam flow rate. V: Secondary Steam Flow rate)

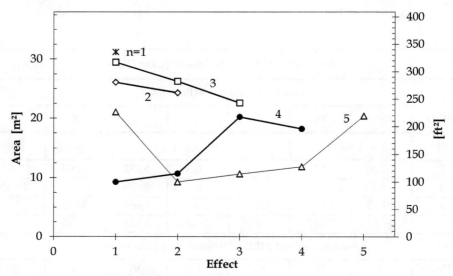

Fig. 25. Area profiles in the 1 to 5-effect MEES.

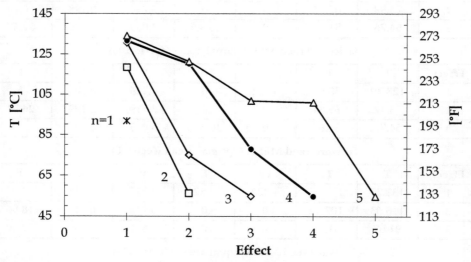

Fig. 26. Temperature profiles in the 1 to 5-effect MEES.

The net result is a drastic decrease in the area of thermal transfer, as you can see through the cost of the effects (Fig. 27). Moreover, the distribution of the heat transfer area curve changes dramatically, from being monotone decreasing for a triple-effect MEES to have a non-monotonic behavior in the Optimal MEES. Presents a maximum in the effect 3 and a minimum in the effect 1, where the product is removed and in which the coefficient U takes its smallest value (Fig. 25). This was achieved by lowering the flow rate of the solvent evaporated in the first two effects, especially in the first (Fig. 28).

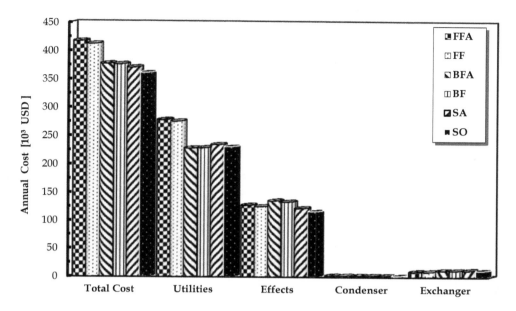

Fig. 27. Impact of the liquid flow pattern and the distribution of heat transfer area in different items of the total annual cost of the MEES (Case III).

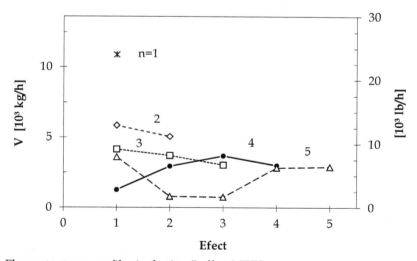

Fig. 28. Flow rate steam profiles in the 1 to 5 effect MEES.

Profile of concentration of the solution in the optimal MESS shows a non-uniform increase of concentration along the evaporator train (Fig. 29). Thus, following the path of the current liquid is seen a percentage increase of 28% in the effect 4, 53% in the effect 3, 73% in effect 2 and 46% in the first effect. Similarly as stated in Case II, the largest concentration jump

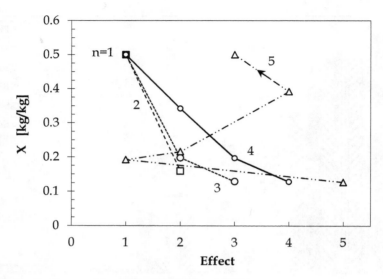

Fig. 29. Concentration profiles in the 1 to 5 effect MEES.

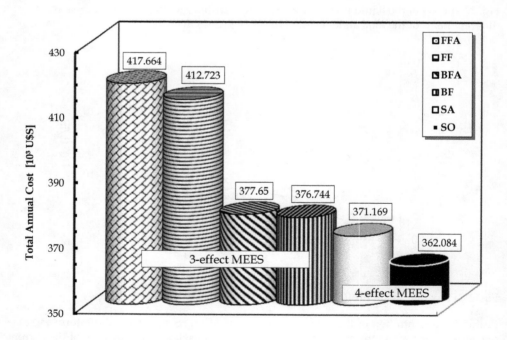

Fig. 30. Impact of the liquid and vapor flow pattern and area effects on the total annual cost of the MEES (Case III).

occurs in effect 2, as opposed to what was observed in Case I, where the largest jump occurs in the last effect, where in addition, the product leaves.

4.3.3 Comparison with traditional configurations

As in the case II, the optimum number of effects of optimal MEES is four and does not match the traditionally proposed by the classic bibliography: triple-effect countercurrent (BFA). If you seek the best solution between traditional structures with effects of equal area (Table 2), again confirms the superiority of the backward feed configuration and that three is the optimal number of effects. Comparatively, FFA is 11.4% more expensive that the structure BFA (Fig. 30). If in both structures allows you to optimize the distribution of the transfer area between the different effects, the improvement is not important. Of them, the best option (BF) is 4.1% more expensive than SO. However, the best configuration for a MEES whose only structural restriction is equality of areas of its effects (SA): is backward feed (BF).

5. Conclusion

In this paper, we solve the problem of designing a MEES, for the concentration of caustic soda, developing a rigorous mathematical model non-convex MINLP type and solving it using a mathematical optimizer.

Unlike previous papers, the new formulation proposal incorporates as decision variables: (a) the trajectories of the steam and liquid flows along the evaporator train whose correct choice determines, in the opinion of different authors, as has been demonstrated in the resolution of the example here presented, the level of operational costs and investment of the MEES. It is also considered (b) the number of effects of the MEES, as was proved in the results presented is another critical design decision and (c) heat transfer area in each effect, without resorting to the hypothesis of equal areas on the effects that in many cases substantially increases the total cost of investment. As an evaluation criterion of alternative designs included in the solutions space of the problem, we used the total annual cost of the system of evaporation, including fixed and operational costs.

Other important aspects, not usually treated and much less simultaneously with the search of the optimal flow pattern, were considered such as:

the rise of the boiling point of the solution and its dependency on temperature and concentration of the same, the variation of the overall heat transfer coefficient along the train following changes in the concentration and the temperature of each effect and the functional dependence of the heat of non-ideal solution with the concentration and temperature.

In addition, we studied other variants non-conventional design that arise by allowing: feed in parallel of the weak solution to two or more effects of the evaporator train, the entry of two or more liquid streams (even of different concentration) to a given effect and the derivation of a liquid solution branch around an effect to avoid its treatment in the same.

A serious drawback found in the resolution of the mathematical model is the presence of many stationary points (optimal local), and also the great influence of the initial point.

It has been found that in order to obtain the optimal design of lower total cost annual: (i) not always steam flow pattern should be the traditional unifilar cascade, could be useful to feed steam live in more than one effect, (ii) the heat transfer areas do not necessarily have to be equal, (iii) the fresh feed stream should not always come in the last effect evaporation train.

However, the synthesis of the optimal MEES only be achieved if you are optimized simultaneously structural, parametric, and operation variables.

6. Acknowledgment

I thank to SeCTER for their help in research.

I thank Liliana, my dear sister, for her love and permanent support throughout my life.

7. References

Foust, Alan S., *Principles of Unit Operations,* Second Ed. C. Wiley, New York, 1980.

Geankoplis, Ch. J., *Transport Processes and Unit Operations,* 2nd. Edition, 495-501, 1983.

Hillebrand Jr., J. B. & Westerberg, A.W., The Synthesis of Multiple-Effect Evaporator Systems Using Minimum Utility Insights - I. A Cascaded Heat Representation *Computers & Chemical Engineering,* Vol. 12, pp. 611-624, 1988.

Horvath, A. L., *Handbook of Aqueous Electrolyte Solutions, Physical Properties, Estimation and Correlation Methods,* Wiley, C. New York, 1985.

Irahola Ferreira, Jaime A., & Jaime Cerdá. Optimal Synthesis Of A Multiple-Effect Evaporator System. *Fifth International Symposium On Process Systems Engineering (PSE),* Korea, 1994.

Irahola Ferreira, Jaime A., & Jaime Cerdá. Síntesis Optima de un Sistema de Evaporación Multiefecto para la Concentración de Licor de Caña de Azúcar. *IX Simposio Internacional en Aplicaciones de Informática. Infonor'96.* Antofagasta. Chile. 1996.

Irahola Ferreira, Jaime A., Aplicación para Formular y Resolver Modelos Matemáticos de Sistemas de Evaporación Multiefecto. *Información Tecnológica.* Chile, 2008

Kern, D. Q. *Process Heat Transfer.* McGraw-Hill, New York .pp 375 - 452, 1999.

Maloney. James O. *Perry's Chemical Engineers Handbook,* 8th Edn. McGraw-Hill, New York 2008.

Nishitani, H. & Kunugita, E., The Optimal Flow-Pattern of Multiple Effect Evaporator Systems, *Computers & Chemical Engineering,* Vol. 3 , pp. 261-268, 1979.

Peters, Max Stone Klaus D. Timmerhaus. *Plantdesign and economics for chemical engineers.* McGraw-Hill, Inc. Singapore, ISBN 0-07-100871-3, 1991.

Standiford, Ferris C. Jr., W. L. Badger, Evaporation. *Chem. Engng,* 70, 158 – 176, 1963.

Westerberg, A.W. & Hillebrand Jr., J. B. The Synthesis of Multiple-Effect Evaporator Systems Using Minimum Utility Insights - II. Liquid Flow-Pattern Selection. *Computers & Chemical Engineering,* Vol. 12, pp. 625-636, 1988.

5

Techno-Economic Evaluation of Large Scale 2.5-Dimethylfuran Production from Fructose

Fábio de Ávila Rodrigues and Reginaldo Guirardello
State University of Campinas, School of Chemical Engineerging
Brazil

1. Introduction

In an era of increasing oil prices and climate concerns, biofuels have gained more and more attention as potential fuel alternative energy sources. Governments have become active in the target of securing a supply of raw materials and limiting climate change, and many innovative proposals have been made, development work has started and potential candidate fuels have been studied in the energy area (Schaub & Vetter, 2008).

A number of factors must be considered when evaluating biofuels: technical factors (raw materials, supply, conversion and engines), economic (engine modification cost, infrastructure) and ecological/political (greenhouse gases, land use efficiency, oil dependence reduction) (Festel, 2008).

An end-user survey assessed car customer acceptance and attitude toward biofuels and revealed that their main demands are: price (48%), biofuel price should not exceed fossil fuels prices and there should be no cost in engine modification; environment (24%); consumption (19%) and performance (9%) (Festel, 2008).

Since customers consider the final cost as a decisive factor, the economic analysis is an important tool in the assessment of the success of biofuel production process and consequent market success. Achieving economic viability used to be the key to success, but today, other factors are important, such as sustainability.

Leshkov et al. (2007) show a catalytic strategy for the production of 2.5 dimethylfuran (DMF) from fructose (a carbohydrate obtained directly from biomass or by the isomerization of glucose) for use as a liquid transportation fuel. Compared to ethanol, 2.5-dimethylfuran has a higher energy density (by 40 percent), a higher boiling point (by 20K), and is not soluble in water. This catalytic strategy creates a route for transforming abundant renewable biomass resources into a liquid fuel suitable or the transportation sector and it is also a CO_2 free process.

The first step in production is to convert fructose to hydroxymethylfurfural (HMF) using an acid catalyst (HCl) and a solvent with a low boiling point in a biphasic reactor. The reactive aqueous phase in the biphasic reactor contains acid and sugar, and the extractive phase contains a partially miscible organic solvent (eg, 1-butanol) that continuously extracts HMF. The addition of a salt to the aqueous phase improves the partitioning of HMF into the extracting phase, and leads to increased HMF yields without the use of high boiling point solvents. Following, water, HCl and solvent evaporate, leading to precipitation of NaCl. Then, HMF is converted into DMF under a copper-ruthenium based catalyst. The final step involves the separation of DMF from the solvent and the non-reacted intermediates. As described below, the process also involves two separation steps. A schematic diagram of fructose conversion to DMF was reported by Leshkov et al. (2007).

The purpose of this paper is to evaluate economically the process production of DMF from fructose. In the present work the following analysis were carried out: Firstly, thermodynamic process modeling was investigated. Following this, the Process Flow Diagram (PFD) was developed from schematic diagram reported by Leshkov et al. (2007). The simulation stage makes use data from Leshkov et al. (2007). The thermal energy required for each piece of equipment was assessed with material and energy balances for each system using the UNISim™ software. Each piece of equipment is then approximately sized for economic analysis.

2. Thermodynamic modeling

The thermodynamic equilibrium of a system consisted of a multicomponent mixture takes place when temperature, pressure and chemical potential of components are equated between the phases, for each component. Although there are other basic criteria for system equilibrium, the minimization of Gibbs free energy is the condition which ensures equilibrium. Salt can affect the solubility of the system components with the formation of complex associations. In general it can be inferred that the particles (molecules, ions, or both) of dissolved salt tend to attract molecules from one of the system components more strongly than others.

The work of Debye and Huckel (1923) was the first important academic contribution and established a model for long-range interactions between ions based on the concept of ionic strength. A different family of models was developed using another extension of the Debye-Huckel model to represent the different contributions to Gibbs free energy excess. Implementation of the local composition to electrolytes means it is governed by local interactions such as short-range solvent/solvent, short-range ion/solvent and long-range ion/ion interactions that exist around the immediate neighborhood of a central ionic species (Aznar, 1996). For the contribution of short-range the following models of local composition can be used: Non-Random Two Liquid model (NRTL) (Renon and Prausnitz, 1968), UNIQUAC (Abrams and Prausnitz, (1975)) or UNIversal Functional Activity Coefficient (UNIFAC) (Fredenslund et al., 1977). The Debye-Huckel term or one of its variations, such as Fowler and Guggnheim (1949) or the Pitzer (1973) are used for long-range interactions. A series of different combinations have been proposed with these elements.

The hypothesis in this work was that salt should be treated as simple molecule, non-dissociated, rather than as charged ions distributed in the solution. Most works concerning the phase equilibrium in systems containing electrolytes distinguish long-range contributions due to electrostatic forces between ions and between ion and solvent from short-range contributions due to interactions between molecules. Two different models are then used for each contribution type. Considering salt as a simple molecule eliminates both contributions and requires only one appropriate model to describe the interactions between all molecules in solution, solvent or electrolyte (Aznar, 1996). Therefore, no specific model for electrolytes was used in this study. The UNIQUAC model was used to obtain the activity coefficient. According to Mock et al. (1986), although the contribution of long-range interaction of the equation of Pitzer-Debye-Huckel is important to obtain the ionic activity coefficient in the aqueous phase, it has little effect on the behavior of the equilibrium phase of the water-organic solvent system. Thus, the effect of the electrolyte is considered only for non-ideality, represented by the adjustable model parameters.

The binary interaction parameters of UNIQUAC model were estimated from experimental data (Santis et al., 1976a, 1976b), from Catté et al. (1994) and for the UNIFAC method. The tables 2, 3 e 4 show the data experimental used to estimate the binary interactions parameters. A Fortran programming language was used to determine the parameters from experimental data. The UNISIM™ software was used to estimate parameters for the UNIFAC method.

	Water	1-butanol	DMF	Fructose	HCl	HMF	NaCl
Water	-	89.22*	1543'	-153.35**	1160'	1361'	-455*
1-butanol	208*	-	383'	223'	1421'	530'	863*
DMF	249'	-90'	-	73'	-	-146'	-371'
Fructose	324**	91'	892'	-	-197"	1.412'	160'
HCl	-674'	-179'	-	399'	-	702'	-266'
HMF	-121'	-1155'	564'	162'	2,776'	-	2479'
NaCl	-165*	1251*	1793'	354'	2943'	1391'	-

* Binary interaction parameters of UNIQUAC model estimated from experimental data from Santis et al. (1976a, 1976b).
** Catté et al. (1994).
' Binary interaction parameter estimated from UNIFAC method.

Table 1. Binary interaction parameters of UNIQUAC model

Aqueous Phase			Organic Phase		
Water	1-butanol	NaCl	Water	1-butanol	NaCl
92.60	7.4	-	20.4	79.6	-
92.04	6.8	1.16	18.78	81.2	0.025
91.64	6.1	2.26	17.45	82.5	0.045
90.85	5.8	3.35	16.64	83.3	0.061
90.60	5.0	4.40	15.43	84.5	0.074
89.96	4.6	5.44	14.6	85.3	0.086
89.16	4.4	6.44	14.1	85.8	0.095
87.84	3.7	8.46	13.29	86.6	0.110
86.30	3.3	10.4	12.48	87.4	0.122
85.10	2.7	12.2	11.37	88.5	0.130
83.50	2.5	14.0	10.66	89.2	0.138
82.20	2.0	15.8	8.75	90.1	0.144
80.90	1.7	17.4	9.05	90.8	0.148
79.40	1.5	19.1	8.55	91.3	0.153
78.00	1.3	20.7	7.94	91.9	0.156
76.70	1.1	22.2	7.34	92.5	0.159
75.20	0.9	23.9	7.04	92.8	0.162
74.00	0.8	25.2	6.54	93.3	0.164
73.30	0.8	25.9	6.43	93.4	0.167
73.30	0.8	25.9	6.23	93.6	0.167

Table 2. Liquid-liquid equilibrium in the system water-1-butanol-NaCl (Santis et al., 1976a)

Aqueous Phase			Organic Phase		
water	1-butanol	NaCl	water	1-butanol	NaCl
92.90	7.10	-	20.60	79.4	-
92.43	6.42	1.15	18.77	81.2	0.026
92.04	5.70	2.26	17.75	82.2	0.045
91.22	5.44	3.34	16.74	83.2	0.061
90.59	5.00	4.41	15.92	84.0	0.075
89.96	4.59	5.45	15.31	84.6	0.086
89.28	4.24	6.48	14.70	85.2	0.096
87.9	3.60	8.50	13.69	86.2	0.111
86.46	3.04	10.5	12.68	87.2	0.123
85.01	2.59	12.4	11.87	88.0	0.132
83.35	2.45	14.2	11.26	88.6	0.140
82.19	1.81	16.0	1.15	89.7	0.146
80.79	1.51	17.7	9.35	90.5	0.150
79.67	1.23	19.1	8,74	91.1	0.155
77.97	1.03	21.0	7.84	92.0	0.158
76.51	0.89	22.6	7.44	92.4	0.161
75.15	0.75	24.1	6.94	92.9	0.164
73.81	0.69	25.5	6.63	93.2	0.166
73.34	0.68	26.0	6.43	93.4	0.169

Table 3. Liquid-liquid equilibrium in the system water-1-butanol-NaCl (Santis et al., 1976a)

Aqueous Phase			Organic Phase		
water	1-butanol	NaCl	water	1-butanol	NaCl
93.40	6.60	-	21.40	78.6	-
92.85	5.96	1.19	19.68	80.3	0.021
92.44	5.25	2.31	18.36	81.6	0.041
91.88	4.70	3.42	17.54	82.4	0.060
90.86	4.61	4.53	16.83	83.1	0.072
90.21	4.21	5.58	16.31	83.6	0.085
89.49	3.86	6.65	15.50	84.4	0.095
88.04	3.27	8.69	14.49	85.4	0.111
86.62	2.78	10.6	13.58	86.3	0.122
85.19	2.31	12.5	12.77	87.1	0.131
83.70	1.90	14.4	11.96	87.9	0.139
82.10	1.70	16.2	11.25	88.6	0.146
80.92	1.28	17.8	10.35	89.5	0.151
79.59	1.01	19.4	9.44	90.4	0.156
78.18	0.82	21.0	8.84	91.0	0.160
76.74	0.66	22.6	8.34	91.5	0.163
75.36	0.54	24.1	7.83	92.0	0.166
73.9	0.50	25.6	7.53	92.3	0.170
73.42	0.48	26.1	7.03	92.8	0.173
73.42	0.48	26.1	7.03	92.8	0.173

Table 4. Liquid-liquid equilibrium in the system water-1-butanol-NaCl (Santis et al., 1976b)

3. Simulation

Simulation of DMF plant production was based on the standard conditions by Leshkov et al. (2007) from which it was developed a process flow diagram (PFD). The following unit operations compose the production plant: pumps, heat exchangers, one reactor for conversion of fructose into HMF (CRV-102) and one reactor for conversion of HMF into DMF (CRV-101), two stripping columns (T-100 and T-101), one distillation column (T-102). The volume of feed was of 30% fructose and the ratio between the organic phase and the aqueous phase volume was of 3.1 in the biphasic reactor (CRV-102). The conversion of fructose is 75% and the conversion of HMF to DMF is 100%.

Unreacted fructose was recycled back into the biphasic reactor. 1-Butanol was then separated from the water in the organic biphasic reactor. Cezário et al. (2009) proposed a separation system for water and 1-butanol composed by two stripping columns, one cooler and one settling tank. The formation of heterogeneous azeotrope turns this separation process more difficult and two liquid phases are formed in the decanter. This system can separate 98% of 1-butanol. Literature provides various processes for separating 1-butanol from water but the most traditional recovery process is distillation. Other techniques are adsorption, liquid-liquid extraction, evaporation and reverse osmosis. The energy required to recover 1-butanol by adsorption is of 1948 kcal/kg while the stripping column method requires 5789 kcal/kg. Other techniques such as perevaporation requires 3295 kcal/kg 1-butanol (Qureshi et al., 2005). The last step was to separate DMF from 1-butanol. The proposed separation system was composed by a distillation column (T-102) which separated 92% of DMF. The T-102 operates with reflux rate of 85 kgmol/h and top component (DMF) fraction of 0.9. The 1-butanol recovered in the T-102 was recycled.

Thus, material and energy balances were then solved using UNISim™ software and is showed in Table 5.

	1h	2h	3h	5h	8
Temperature (°C)	25	25	25	25	25
Pressure (kPa)	101	101	101	101	1351
Massic flow (kg/h)	896	1000	864	52	896
Enthalpy (kJ/kgmol)	-275200	-290600	-290500	-326600	-275200
Composition (massic fraction)					
water	0.9	0.7	0.5372	-	0.9
1-butanol	-	-	-	1	-
Fructose	-	0.3	-	-	-
Hmf	-	-	-	-	-
NaCl	-	-	0.4628	-	-
HCl	0.1	-	-	-	0.1
NaOH	-	-	-	-	-
DMF	-	-	-	-	-

	9	10	11	12	13
Temperature (°C)	25	25	180	180	180

Pressure (kPa)	1351	101	1351	1351	1351
Massic flow (kg/h)	1000	864	4142	896	1000
Enthalpy (kJ/kgmol)	-290600	-290500	-326600	-263000	-278700
Composition (massic fraction)					
Água	0.7	0.5372	-	0.9	0.9589
1-butanol	-	-	1	-	-
Frutose	0.3	-	-	-	0.0411
Hmf	-	-	-	-	-
NaCl	-	0.4628	-	-	-
HCl	-	-	-	0.1	-
NaOH	-	-	-	-	-
DMF	-	-	-	-	-

	14	16	15	18	19
Temperature (°C)	180	180	180	180	37
Pressure (kPa)	1351	1351	1351	1351	1351
Massic flow (kg/h)	864	7392	489	7392	7392
Enthalpy (kJ/kgmol)	-260600	-291400	-274800	-277200	-294200
Composition (massic fraction)					
water	0.7901	0.3250	0.8825	0.3250	0.3250
1-butanol	-	0.5603	-	0.5603	0.5603
Fructose	-	0.0484	0.1174	0.0121	0.0121
Hmf	-	-	-	0.0363	0.0363
NaCl	0.2098	0.0541	-	0.0541	0.0541 .
HCl	-	0.0122	-	0.0218	0.0218
NaOH	-	-	-	0	0
DMF	-	0.00005	-	0	0

	22	29	30	25	24
Temperature (°C)	37	37	37	80	81
Pressure (kPa)	1351	1351	1351	50	50
Massic flow (kg/h)	7592	4876	2716	23575	489
Enthalpy (kJ/kgmol)	-296500	-311400		-240600	-282400
Composition (massic fraction)					
water	0.3353	0.0887	0.7781	0.7995	0.8822
1-butanol	0.5456	0.8444	0.0091	0.1876	-
Fructose	0.0118	0.0118	0.01168	-	0.1178
Hmf	0.0353	0.0549	-	0.01138	-
NaCl	0.0719	-	0.2011	-	-
HCl	-	-	-	-	-
NaOH	-	-	-	-	-
DMF	-	-	-	0.0015	-

	26	27	Organic phase	Aqueous phase	32

Temperature (°C)	74	71	36	36	71
Pressure (kPa)	50	50	50	50	50
Massic flow (kg/h)	78804	55229	59614	19189	55230
Enthalpy (kJ/kgmol)	-224500	-197800	-245800	-284100	-197800
Composition (massic fraction)					
water	0.2585	0.0275	0.0255	0.4822	0.02752
1-butanol	0.1716	0.1647	0.2217	0.01592	0.1647
Fructose	-	-	-	-	-
Hmf	0.00341	-	0.004517	-	-
NaCl	-	-	-	-	-
HCl	-	-	-	-	-
NaOH	-	-	-	-	-
DMF	0.5665	0.8077	0.7483	0.00177	0.8077

	33	34	35	38	39
Temperature (°C)	100	101	219	71	99
Pressure (kPa)	50	1650	1650	50	50
Massic flow (kg/h)	4384	4384	4384	290	4094
Enthalpy (kJ/kgmol)	-390900	-308800	-279700	-215100	-310900
Composition (massic fraction)					
water	-	-	-	-	-
1-butanol	0.9388	0.9388	0.9388	0.07891	0.9999
Fructose	-	-	-	-	-
Hmf	0.0611	0.0611	0.0611	-	-
NaCl	-	-	-	-	-
HCl	-	-	-	-	-
NaOH	-	-	-	-	-
DMF	-	-	-	0.9221	0.000086

	40
Temperature (°C)	-
Pressure (kPa)	0.9388
Massic flow (kg/h)	-
Enthalpy (kJ/kgmol)	0.0611
Composition (massic fraction)	
water	-
1-butanol	-
Fructose	-
Hmf	-
NaCl	-
HCl	0.9388
NaOH	-
DMF	0.0611

Table 5.

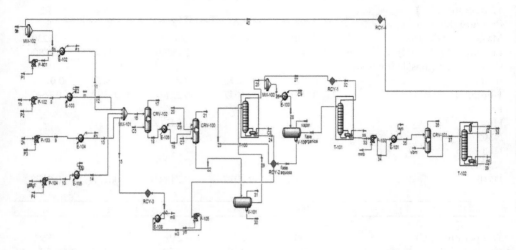

Fig. 1. Material and energy balance for each stream in DMF production plant.

4. Economic evaluation

The economic evaluation was based on the spreadsheets by Peters & Timmethaus (2003). The following steps were used by performed the economic analysis.

i.   On the sheet 'Capital Inv.' The estimated current total purchased cost of the equipament was entered. The results are showed in Table 7.
ii. On the sheet 'Materials & Labor' the product prices and flowrates, the raw materials prices and flow rates, and the labor requirements were entered. The results are showed in Table 8.
iii. On the sheet 'Utilities' the quantity of each utility needed annually was entered in appropriate units. The total annual utilities cost is transferred to sheet 'Annual TPC';
iv. The 'Depreciation' sheet is used only if the user wishes to change the default (5-year Modified Accelerated Cost Recovery System (MACRS) depreciation method);
v.   On the 'Annual TPC' sheet, all values were calculated from information available on other sheets. The Calculated annual TPC was transferred to 'Evaluation'. The results are showed in Table 9.
vi. The sheet 'Evaluation' used values from other sheets to calculate the common profitability measures. All calculations in 'Evaluation' are made in current (i.e. inflated) dollars.

Each piece of equipment was roughly sized from material and energy balance and the approximate cost determined. Costs of equipment operating at ambient pressure and using carbon steel, were estimated by Eq. (1) (Turton et al., 2003).

$$logCp^o = K_1 + K_2 log(A) + K_3 (log(A))^2 \qquad (1)$$

Where A is the equipmen t capacity or size parameter and K_1, K_2 and K_3 are the parameters (Turton et al., 2003). The effect of time, operating conditions and material construction on

Equipment number	equipment	Parameter equipment	Equipment cost ($), CEPCI = 499,6
P-101	pump	flow: 4.97 m³/h	3814
P-102	pump	flow: 0.92 m³/h	2910
P-103	pump	flow 0.864 m³/h	2898
P-104	pump	flow: 0.63 m³/h	2845
P-105	pump	flow: 0.54 m³/h	2823
P-100	pump	Flow: 5.27 m³/h	4670
E-102	heater	Heaty duty: 302, 5 KW	64862
E-103	heater	Heaty duty: 160 KW	42107
E-104	heater	Heaty duty: 134 KW	37333
E-105	heater	Heaty duty: 107 KW	320119
E-106	heater	Heaty duty: 8790 KW	637790
E-108	heater	Heaty duty: 100 KW	30610
E-100	cooler	Water flow: 198 m³/h	11421
E-101	heater	Heaty duty:467 KW	87085
CRV-102	reactor	Heaty duty: 10000 KW	2917722
CRV-100	reactor	Heaty duty: 344,75 KW	22354
CRV-101	reactor	Heaty duty: 10000 KW	2917722
T-100	stripping colunn	Height: 4 m; diameter: 1m	15003
T-101	stripping colunn	Height: 4 m; diameter: 1m	15003
T-102	destilation colunn	Height: 4 m; diameter: 1m	15003
Total			1609365

Table 6. Equipment parameter

	Fraction of delivered equipment			% chosed (B)	Calculated values, million ($)
	Solid processing plant (A)	Solid fluid processing plant (B)	Fluid processing plant (C)		
Cost directs					
Purchased equipment, E'					1.609
Delivery, fraction of E'	0.10	0.10	0.10	0.10	0.1604
Purchased equipment instalation	0.45	0.39	0.47	0.39	0.6903
Instrumentation & controls (installed)	0.18	0.26	0.36	0.26	0.460
Piping (installed)	0.16	0.31	0.68	0.31	0.548
Eletrical systems (installed)	0.10	0.10	0.11	0.10	0.177

	Fraction of delivered equipment			% chosed (B)	Calculated values, million ($)
	Solid processing plant (A)	Solid fluid processing plant (B)	Fluid processing plant (C)		
Buildings (incluing services)	0.25	0.29	0.18	0.29	0.513
Yard improvements	0.15	0.12	0.10	0.12	0.212
Service facilities (installed)	0.40	0.55	0.70	0.55	0.973
Total direct costs					5.345
Indirect costs					
Engineering and supervision	0.33	0.32	0.33	0.32	0.566
Construction expenses	0.39	0.34	0.41	0.34	0.602
Legal expenses	0.04	0.04	0.04	0.04	0.071
Contractor's fee	017	0.19	0.22	0.19	0.336
contigency	0.35	0.37	0.44	0.37	0.655
Total indirect costs					2.23
Fixed capital investimento					7.575
Working capital	0.70	0.75	0.89	0.75	1.327
Total capital investment					8.903

Table 7. Estimation of capital investment by percentage of delivered equipment method

Material	classification	Price (US$/kg)	Annual amount (million kg/year)	Annual value (million US$/year)
DMF	Product	variable	2.217	0.88
Fructose	Raw material	0.78 (variable)	2.484	1.94(62%)
NaCl	Raw material	0.015	0.05	0.0007(0.02%)
HCl	Raw material	0.295	0.745	0.22(7.7%)
1-butanol	Raw material	1.72	0.431	0.7906(23%)
water	Raw material	0.08	16.312	1.30(2.6%)
NaOH	Raw material	0.10	25	2.5(2.7%)
H_2	Raw material	10	0.13834	1.3834(1.4%)

Table 8. Annual raw material costs and products values

Item (A)	Factor (B)	Basis (C)	Basis Cost (million) US$/year (D)	Cost (milion) US$/year (E)
1. Raw materials				4.204
2. operating labor (M)				0.885
3. operating supervision	0.15	de (2)	0.885	0.133
4. utilities				0.55
5. maintenance and repairs (MR)	0.06	de FCI	1.407	0.46
6. operating supplies	0.15	de (5)	0.084	0.07
7. laboratory charges	0.15	de (2)	0.885	0.133
8. Royalties	0.01	de c_o	3.090	0.08
9. catalysts and solvents	0			2.5
Variable cost				6.536
10. taxes (property)	0.02	de FCI	1.407	0.156
11. financing (interest)	0	de FCI	1.407	0
12. insurance	0.01	de FCI	1.407	0.014
13. rent	0		1.407	0.078
14. depreciation		Calculated separately		
C				0.233
15. plant overhead, general	0.6		1.102	0.891
16. administration				0.661
17. manufacturing cost				7.660
18. administration	0.2	de (2), (5)	1.102	0.297
19. distribution & selling	0.05	de c_o	3.090	0.437
20. Research & development	0.04	de c_o	3.090	0.350
General expense				0.469
Total product cost without depreciation = c_o				8.744

Table 9. Annual total product cost at 100 % capacity

purchased equipment cost was corrected by time factor (I), material factor (F_M) and conditions factor (F_P). Purchased equipment cost is then expressed by:

$$Cp = Cp^0 \times F_M \times F_P \times I \qquad (2)$$

In this work, inflation was account ed for by the Chemical Engineering Plant Cost Index (Lozowski, 2010). According the Table 8 the price of raw material and solvent are the more expensive.

With chemical and utility cost were obtained and a discounted cash flow analysis was performed to determine profitability. The quantities of chemical material, utilities and production of DMF were doubled, tripled, etc, from the simulated plant, to achieve sale price and cost DMF similar to gasoline and ethanol. However, the equipment cost increased according to Eq. (3). For all sale price and cost DMF from Tables 5, 6 e 7 the profitability measures were: 15.0 %/year (return on investment) and a 3.6 year payback period.

The conversion and the price fructose were changed too. The tables 10, 11 and 12 show the results.

	Conversion (75%)	Conversion (80%)	Conversion (85%)
Fructose price, 0.78 US$/kg	6.00/3.94	5.90/3.83	5.80/3.75
Fructose price, 0.50 US$/kg	5.70/3.59	5.60/3.49	5.40/3.42
Fructose price, 0.10 US$/kg	5.20/3.10	5.00/3.00	4.90/2.94

Table 10. Sale price of DMF/cost (US $/kg) of DMF to several conversions in reactor CRV-102 and various prices of fructose (standard plan used in the simulation).

	Conversion (75%)	Conversion (80%)	Conversion (85%)
Fructose price, 0.78 US$/kg	3.70/2.68	3.60/2.59	3.52/2.54
Fructose price, 0.50 US$/kg	3.30/2.33	3.20/2.26	3.13/2.22
Fructose price, 0.10 US$/kg	2.70/1.83	2.63/1.78	2.58/1.74

Table 11. Sale price of DMF/cost (US $/kg) of DMF to several conversions in reactor CRV-102 and various prices of fructose (scale factor 12).

	Conversion (75%)	Conversion (80%)	Conversion (85%)
Fructose price, 0.78 US$/kg	2.68/1.95	2.60/1.89	2.55/1.85
Fructose price, 0.50 US$/kg	2.27/1.60	2.20/1.56	2.15/1.52
Fructose price, 0.10 US$/kg	1.68/1.10	1.63/1.07	1.60/1.05

Table 12. Sale price of DMF/cost (US $/kg) of DMF to several conversions in reactor CRV-102 and various prices of fructose (scale factor 30).

In table 12 observes that the sale price of DMF can be compared with the gasoline. The cost of DMF decreases with the increase of the conversion of fructose to HMF and with the price decrease of fructose.

$$Equipament\ cost = N^{0.6}$$ (3)

Where N is the scale factor with values of N = 2, 5, 10, etc.

Thus, the estimated cost of the equipment was U.S. $ 12 million, fixed capital investment was U.S. $58 million, direct cost were U.S. $41 million, indirect costs were U.S. $17 million, working capital was U.S. $10 million and total capital investment was U.S. $68 million. From economic evaluation the value and cost DMF was 2.68 U.S. $/kg and 1.95 U.S $/kg, respectively. For this analysis, the plant is economically feasible for a scale factor of thirty (N= 30).

5. Conclusions

The following conclusions can be drawn from the facts presented in the above review. In the thermodynamic analysis salt is considered a solute, so it´s possible to use the model UNIQUAC (Mock et al., 1986). The estimation of binary interaction parameters for UNIQUAC in the system water-butanol-salt was carried out with Fortran software from liquid-liquid equilibrium data and UNIFAC (UNIQUAC Functional-group Activity Coefficient) method was used to estimate remain parameters. The separation system (composed by two stripping columns, one cooler and one settling tank) used to separate 1-butanol and water recovery 98% of 1-butanol. The separation system (composed by distillation column) used to separate DMF recovery 92 % of DMF. Economic evaluation showed that a suitable operational plant could work with 12.4 tons/year of fructose. It could produce 11.1 tons/year of DMF. The fixed capital investment in plant and equipment is estimated at U.S. $ 58 million and U.S. $ 12 million, respectively. The DMF value was 2.69 U.$. $/kg. For this analysis, the plant is economically feasible, from comparison with a reference market of 15.0 %/year (return on investment) with a 3.6 year payback period. This analysis suggests that DMF production from fructose deserves serious consideration by investors.

6. References

Abrams, D.S. & Prausnitz, J.M. (1975). Statistical Thermodynamics of Liquid Mixtures: A New Expression for the Excess Gibbs Energy of Partly or Completeley Miscible Systems. *American Institute of Chemical Engineers Journal*, Vol.21, pp. 116-128.

Allen, D.H. (1980). *A Guide to Economic Evaluation of Projects*, The Institutions Chemical Engineers, Rugby, United Kingston.

Aznar, M. (1996). *Equilíbrio Líquido-Vapor de Sistemas com Eletrólitos Via Contribuição de Grupo*, Universidade Federal do Rio de Janeiro, Rio de Janeiro, Brazil.

Catté, M.; Dussap, C.G.; Achard, C. & Gros, J.B. (1994). Excess Properties and Solid-Liquid Equilibria for Aqueous Solutions of Sugars Using a UNIQUAC Model. *Fluid Phase Equilibria*, Vol.96, pp. 33-50.

Cezário, G. L.; M. Filho, R. & Mariano, A. P. (2010). Projeto e Avaliação Energética do Sistema de Destilação de uma Planta de Fermentação Extrativa para a Produção de Biobutanol, *Proceedings of XVII Congresso de Iniciação Científica da Universidade Estadual de Campinas*, Campinas, Brazil.

Debye, P. & Huckel, E. (1923). Zur Theorie der Elektrolyte. *Physics Zeitsch*, Vol.24, pp. 185-206.

Festel, G. W. (2008). Biofuels - Economic Aspects. *Chemical Engineering Technology*, Vol.31, pp. 715-720.

Fowler, R.H. & Guggenheim, E.A. (1949). *Statistical Thermodynamics*, Cambridge University Press, Cambridge.

Fredenslund, A.A.; Gmehling, J. & Rasmussen, P. (1977). *Vapor-Liquid Using UNIFAC*, Elsevier, Amsterdam.

Leshkov, Y.R.; Barrett, C. J.; Liu, Z.Y. & Dumesic, J.A. (2007). Production of Dimethylfuran for Liquid Fuels from Biomass-Derived Carbohydrates. *Nature*, Vol.447, pp. 982-986.

Lozowski, D. (2010). Economic Indicators. *Chemical Engineering*, Vol.117, pp. 55.

Mock, B.; Evans, L.B. & Chen, C.C. (1986). Thermodynamic Representation of Phase Equilibria of Mixed-Solvent Electrolyte Systems. *Association International of Chemical Engineers Journal*, Vol.32, pp. 1655-1664.

Peters, M.S.; Timmerhaus, K.D. & West, R.E.W. (2003), *Plant Design and Economics for Chemical Engineers*, McGraw-Hill, New York.

Pitzer, K.S. (1973). Thermodynamics of Electrolytes I. Theoretical Basis and General Equation. *Journal Physics Chemical*, Vol.77, pp. 268-277.

Qureshi, N.; Hughes, S.; Maddon, I.S. & Cotta, M.A. (2005). Energy-Efficient Recovery of Butanol from Model Solutions and Fermentation Broth by Adsorption. *Bioprocess and Biosystems Engineering*, Vol.27, pp. 215-222.

Renon, H. & Prausnitz, J. M. (1968). Local Compositions in Thermodynamics Excess Functions for Liquid Mixtures. *American Institute of Chemical Engineers Journal*, Vol.14, pp. 135-144.

Turton, R.; Bailie, R.C.; Whiting, W.B. & Shauwitz, J.A. (2003), *Analysis, Synthesis, and Design of Chemical Processes*, Prentice Hall, New Jersey.

Santis, R.; Marrelli, L. & Muscetta, P.N. (1976a). Liquid-Liquid Equilibria in Water-Aliphatic Alcohol Systems in the Presence of Sodium Chloride. *Chemical Engineering Journal*, Vol.11, pp. 207-214.

Santis, R.; Marrelli, L. & Muscetta, P.N. (1976b). Influence of Temperature on the Liquid-Liquid Equilibrium of the Water-n Butyl Alcohol-Sodium Chloride System. *Journal of Chemical and Engineering Data*, Vol.21, pp. 324-327.

Schaub, G. & Vetter, A. (2003). Biofuels for Automobiles - An Overview. *Chemical Engineering Technology*, Vol.31, pp. 721-729.

UNISim Honeywell (2007), http://www.honeywell.com.

6

Inland Desalination: Potentials and Challenges

Khaled Elsaid[1], Nasr Bensalah[1,2,*] and Ahmed Abdel-Wahab[1]

[1]Department of Chemical Engineering,
Texas A&M University at Qatar,
Education City, Doha
[2]Department of Chemistry, Faculty of Sciences of Gabes,
University of Gabes, Gabes
[1]Qatar
[2]Tunisia

1. Introduction

Groundwater is the main source of drinking water in many countries all over the world. In absence of surface water supply, the use of groundwater as the main water source for drinking, industrial, and agricultural use becomes essential especially in the case of rural communities. Underground reservoirs constitute a major source of fresh water, in terms of storage capacity; underground aquifers worldwide contain over 95% of the total fresh water available for human use. Typical groundwater supplies have low coliform counts and total bacterial counts, low turbidity, clear color, pleasant taste, and low odor. Accordingly, groundwater has higher quality than surface water, and the quality is quite uniform throughout the year that makes it easy to treat. A disadvantage of groundwater supplies is that many groundwater aquifers have moderate to high dissolved solids such as calcium, magnesium, iron, sulfate, sodium, chloride, and silica. The high concentration of dissolved solids particularly, sodium chloride, makes the water brackish and thus requires to be desalinated before its use for a certain purpose.

With the growth of membrane science, reverse osmosis RO overtook multi stage distillation MSF as the leading desalination technology. In the last two decades, RO processes have advanced significantly, allowing new brackish groundwater desalination facilities to use RO technology much more economically than distillation. RO treatment plants use semipermeable membranes and pressure to separate salts from water. These systems typically use less energy than thermal distillation, leading to a reduction in overall desalination costs.

The reverse osmosis process enables now the massive production of water with a moderate cost, providing flexible solutions to different necessities within the fields of population supply, industry and agriculture. The great development of reverse osmosis (RO) technology has been a consequence of several factors such as reduction in energy consumption and membrane cost. Nevertheless, the major problem of RO desalination

---

* Corresponding Author

plants is the generation of a concentrate effluent (brine) that must be properly managed. Disposal of such brines presents significant costs and challenges for the desalination industry due to high cost and environmental impact of brine disposal.

The reject brine from desalination plants not only contains various types of salts at higher concentration but also other types of wastes like pretreatment chemicals (antiscalents, antifouling,...etc). Also, if the feed water includes harmful chemicals such as heavy metals or others, these chemicals are concentrated in the reject brine. Improper disposal of reject brine from inland plants results in several environmental problems.

Although sea disposal of reject brine is a common practice for plants located in coastal areas, it would not be available for inland desalination. Deep well injection is prohibitly expensive and has its own problems such as the possibility of corrosion and subsequent leakage in the well casing, seismic activity which could cause damage to the well and subsequently result in ground water contamination, and uncertainty of the well life. Additionally, when a sewerage system is used for disposal of concentrate brine high in total dissolved solids (TDS) the treated municipal sewage effluent becomes unsuitable for reuse.

While operation and maintenance costs for evaporation ponds are minimal, large land areas are required, and pond construction costs are high. Even in arid climates ideally suited for evaporation, a typical design application rate is only 2 gpm per acre. The construction cost for an evaporation pond with a liner and monitoring system typically ranges between $100,000 and $200,000 per acre, exclusive of land cost. Thus a concentrate flow as small as 100 gpm would require a pond area of at least 50 acres and cost $10 to $20 million to construct. Consequently, evaporation ponds are often cost prohibitive and impractical for handling any significant concentrate flow. Furthermore, water evaporated from a pond is often a lost resource.

Therefore, the need to protect surface and groundwater resources may in many cases preclude concentrate disposal by the earlier three methods. The alternative is zero liquid discharge (ZLD). In ZLD, concentrate is treated to produce desalinated water and essentially dry salts. Consequently there is no discharge of liquid waste from the process. Most ZLD applications in operation today treat industrial wastewater or power plant cooling water using thermal crystallization, evaporation ponds, or a combination of these technologies. Thermal crystallization is energy-intensive with high capital and operating costs.

Given the need for ZLD and the disadvantages of existing ZLD methods, it is imperative to find alternative ZLD treatment technologies that provide more affordable concentrate management. This article reviews current trends and potential advancements of inland desalination and brine management alternatives.

2. Water and water resources

Water resources present naturally in the environment can be generally divided into freshwater and saline water according to the amount of dissolved solids it contains. Quality and quantity of different water resources are of high importance, many efforts are being made to have good estimates of water resources at both worldwide and country levels. In 1990s "The comprehensive assessment of the freshwater resources of the world" was launched to have estimates on worldwide water resources (United Nations [UN], 1997). The

AQUASTAT program (Food and Agriculture Organization [FAO]) was launched to form global information system on water and agriculture, the main objective of the program is to collect and analyze information on water resources, water uses, and agriculture water management within different countries. Information on the quantity of major water resources is present in table 2.1.

Water Resource	Volume, (1000 km³)	Percent of total water	Percent of total fresh water
Saline water:			
Oceans/seas	1,338,000	96.54	-
Saline/brackish groundwater	12,870	0.93	-
Saltwater lakes	85	0.006	-
Freshwater:			
Glaciers and permanent snow covers	24,064	1.74	68.70
Fresh groundwater	10,530	0.76	30.06
Fresh lakes	91	0.007	0.26
Wetlands	11.5	0.001	0.03
Rivers	2.12	0.0001	0.006

Table 1. Estimates of major water resources on Earth (Gleick, 2001).

Freshwater is the water naturally found on Earth's surface and in underground aquifers such as surface water, fresh groundwater, and glaciers, and mainly characterized by its low content of dissolved solids. These water sources are considered to be renewable resource, by effect of natural water cycle. The quantity of freshwater present on Earth is around 2.5% only of the total water present on Earth.

Surface water is the water present in rivers, fresh lakes, and wetlands; the main source of surface water is by precipitation in the form of rain, snow....etc. Surface water is characterized by low content of dissolved salts generally below 500 mg/L. Surface water represents only around 0.3 % of the total freshwater present on Earth's surface. Fresh groundwater is the water located under the Earth's surface i.e. subsurface water which is mainly located in pores or spaces of soil and rocks, or in aquifers below the water table. It is mainly characterized by its low suspended solids. In many places groundwater contains high content of dissolved salts compared to that of surface water; with salinity level around 500-2,000 mg/L (Mickley, 2001). Groundwater represents around 0.76 % of the total water present on Earth, and around 30 % of the freshwater available on Earth.

Water or ice present in glaciers, icebergs, and icecaps represents the vast majority freshwater, this huge amount of water is currently unused and locked up in southern and northern poles. Up to date there is no efforts has been made to make use of such water resources due to the high cost associated with its processing as it is mainly present in very distant areas or at very high altitudes.

Brackish groundwater is the water located under the Earth's surface and it is characterized by its higher salinity than that of fresh groundwater with values of 2,000-10,000 mg/L

(Mickley, 2001). It is mainly present in aquifers that are much deeper than that of fresh groundwater. Brackish groundwater represents around 0.93 % of the total water present on Earth.

Saline or salty water is the water that contains considerable amount of salts and it is mainly found in oceans, seas, saline or brackish groundwater, and saltwater lakes. Saline water represents the majority of water resources in terms of quantity with around 97.5 % of the total water present on Earth. the salinity of seawater varying from one location to another, from around 21 g/L in the North Sea to 40 - 45 g/L in the Arabian Gulf and Red Sea, and even up to 300 g/L as in the Dead Sea (Gleick, 2006).

The majority of world population use surface water or groundwater as the main source for domestic, agriculture, and industrial water supplies. The most common surface water sources are rivers, and lakes. However, the most common groundwater sources are pumped wells or flowing artesian wells. In absence of surface water supply, it is clear that the use of groundwater becomes essential especially in the case of rural communities. Underground reservoirs constitute a major source of fresh water, in terms of storage capacity; underground aquifers worldwide contain over 95% of the total fresh water available for human use. In addition when looking to the map of worldwide water stress in figure 2.2, we find that areas that face water stresses are increasing with time with Middle East, North Africa, and Central Asia having the highest water stresses, while when looking to the worldwide groundwater resources map as shown in figure 2.3, we find that most of these areas have access to groundwater resources, which means that groundwater will present the relief to the faced water stress problems, and even can support the different developmental planes of these areas.

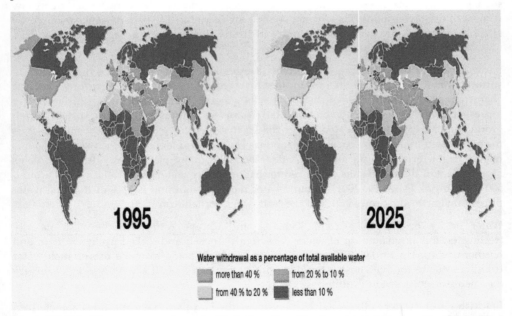

Fig. 1. Worldwide water stresses map (United nations Environmental Programme [UNEP]).

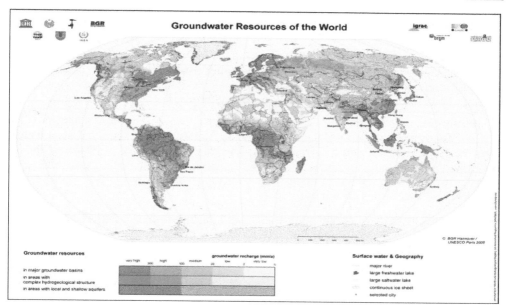

Fig. 2. Worldwide groundwater resources mapping (The Federal Institute for Geosciences and Natural Resources).

3. Groundwater quality

Pure water is a colorless, odorless, and tasteless liquid, water is polar and strong solvent capable of dissolving many natural and synthetic substances, many inorganic and organic compounds, in addition it is able to suspend many solids, and hence it is very hard to find pure water in nature.

The quality of groundwater is mainly determined by its content of dissolved solids and gases, presence of suspended solids, and bacteria present. Usually the nature and concentration of dissolved solids present in groundwater source will depend on characteristics of the aquifer and on travelling time or velocity of groundwater flow through the rock formation (Delleur, 2007).

The physical properties of water are mainly, the total, suspended, and dissolved solids, in addition to turbidity, temperature, color, odor and taste. Typical groundwaters, as they undergo natural filtration while passing through sand formations have very low suspended solids content, low turbidity, clear color, pleasant taste, and low odor. However, while water is travelling through the soil formations, groundwater may carry dissolved solids if the soil formations are relatively soluble. The type and concentration of dissolved solids released from soil to groundwater may vary based on the soil composition, travelling time and flow velocities.

Chemical characteristics are mainly concerned with pH value, cations and anions, alkalinity, acidity, hardness, dissolved gases, and other contaminants such as organic substances and heavy metals that might be present in water. Groundwaters in general have higher hardness when compared to surface waters; this is mainly due to the dissolution of limestone and dolomite formations which in turn increase the content of calcium ions and hence increasing the hardness of water.

Biological characteristics are concerned with the living organisms present in water, mainly bacteria, fungi, algae, and viruses. Microbes are generally absent from groundwater due to natural filtration, and hence groundwater has low coliform and total bacterial counts.

In general groundwater has higher quality than surface waters, and the quality is quite uniform throughout the year making it easy to treat. A disadvantage of groundwater is that many have moderate to high dissolved solids. The high concentration of dissolved solids, particularly sodium chloride, makes the water brackish. The removal of such dissolved solids requires use of desalination process to treat the water to a level of dissolved solids that makes it suitable for a certain use.

Technical and economical evaluations of desalinating groundwater has started early in 1960s. One of the first major brackish groundwater desalination plants was built in Florida, USA with capacity of 2.62 mgd, followed by another one using Reverse Osmosis RO in mid-1970s with capacity of 0.5 mgd (Bart Weiss, 2002). A combination of both field investigations and computer simulation modeling were used to assess the economic suitability of using highly brackish groundwater for large scale abstraction for feeding reverse osmosis desalination plant in many countries around the world (Brich et al., 1985; Hadi, 2002; Sherif et al., 2011; Zubari, 2003). The results from such technical and economical evaluations have shown that brackish groundwater can be considered as high quality source of feedwater for desalination plants, even at higher salinities, which still far below the salinity of seawater. However the evaluations also indicated that the groundwater quality is changing and not constant over long years with continuous increase in dissolved solids content, which should be considered during design stages of such desalination plants.

It was concluded that the major misconception of considering that the groundwater quality is relatively stable over time is very critical for design purposes. The groundwater quality tends to show increase in salinity with time, and hence the initial and future groundwater qualities should be considered (Missimer, 1994). The deteriorating quality of groundwater is becoming of high concern globally, mainly due to human activities such as over abstraction and seawater intrusion. The seawater intrusion is a very common problem worldwide especially, near coastal areas, it is a natural phenomenon in which saline water from sea or ocean moves into the fresh groundwater in coastal aquifers. This behavior is mainly attributed to the density difference and to tidal effects. Seawater intrusion was found to be the main source for the increased salinity of near-coast aquifers in many places worldwide (Amer et al., 2008; Zubari et al., 1994).

4. Inland desalination processes

Desalination is a water treatment process for removing total dissolved solids (TDS) from water. Desalination of seawater and brackish water has become a reliable method for water supply all-over the world and had been practiced successfully for many decades.

The worldwide desalination capacity increased dramatically from around 35 million m^3/d in 2005 (Gleick, 2006) to about 60 million m^3/d by 2009, with the largest desalination plant of 880,000 m³/d at Shoaiba 3 project in Saudi Arabia (International Desalination Association [IDA], 2009). By 2009 there were 14,451 desalination plants online, with further 244 known to be on their way, being under contract or under construction, with an additional capacity of 9.1 million m³/d in 130 countries around the world (IDA, 2009). Desalination in general can be mainly classified according to the feed water source into two main classes as seawater desalination and brackish groundwater desalination. In 2006 around 56 % of the world desalination capacity was for seawater desalination, and 24 % for brackish water, with Saudi Arabia KSA, Unite states US, United Arab Emirates UAE, Spain, Kuwait, and Japan having desalination capacity over 1 million m³/d, with Middle East countries holding over 50 % of the worldwide desalination capacity (Gleick, 2006).

Desalination processes are many and can be generally classified according to the technology used and it mainly classified to thermal desalination, and membrane desalination, ion exchange, and electrodialysis. In addition, other new technologies are still under development. The common processes for desalination have been changed from expensive techniques with extensive energy requirements to a sustainable method for drinking water supply.

The selection of desalination process for a certain purpose depends on many factors such as energy cost, final water quality, fouling propensity, temperature, and overall process cost. Table 2 shows the worldwide, US, and Gulf Corporation Council GCC desalination capacity by processes.

Process Type	Worldwide	United States	GCC Countries
RO	46	69	28
MSF	36	1	54
ED	5	9	-
VC	5	3	-
MED	3	1	9
NF	-	15	-
Others	5	2	9

Table 2. Desalination capacity percentage distribution according to process type (Gleick, 2006).

The quality of desalinated water differs depending on the process used, with thermal desalination producing very high quality product water with salinity about 2-10 mg/L; which usually requires remineralization in the post treatment step (Gabbrielli, 1981). However for membrane process the quality of product water depends on many factors such as the quality of feedwater, design recovery, and membrane properties, but in general the product water will have higher salinity than that produced from thermal desalination.

The quality of the final water product mainly depends on the application in which it is to be used for, ranging from very high quality for process water to specified quality as per the regulation for drinking water, or to a certain quality suitable for agriculture. Most of the desalination plants worldwide are designed to be able to produce high quality water. Drinking water standards vary slightly from country to another. The most widely used is the United States Environmental Protection Agency US-EPA drinking water standards with

500 mg/L for total dissolved solids, and 250 mg/L for each of sulfate and chloride(US EPA, 2009). However the World Health Organization WHO guidelines for drinking water quality suggest that total dissolved solids value of 600-1,000 mg/L is generally acceptable (WHO, 2011). On the other hand the FAO suggests that a total dissolved solids up to 2,000 mg/L is acceptable for irrigation purposes (Ayers & Westcot, 1994).

4.1 Thermal desalination processes

In desalting operations, thermal technologies were the only viable option for long time, and Multi-Stage Flash Distillation (MSF) was established as the baseline technology; however Multiple-Effect Distillation (MED) is now the state-of-the-art thermal technology, but has not been widely implemented yet. Thermal desalination plants have provided the major portion of the world's desalination capacity, as the world's requirements for treated water increase. Thermal desalination is usually used for cases where high salinity feed waters are used i.e. seawater, high recoveries are required, high feedwater temperature, and low energy cost, however the main drawback is the extensive energy consumption (Greenlee et al., 2009).

Thermal desalination or some time called phase change desalination; is a very basic process in concept as it copies the natural process of water cycle, where energy in thermal form through the solar radiation evaporates the water into water vapor, which is condensed later and fall in the form of rains or snow (Gleick, 2001), so in thermal desalination thermal energy or heat is applied to the water present in boiler or evaporator to drive water evaporation, this water vapor is condensed later in condenser by exchanging heat, thus sometimes the process is called phase change desalination as water phase changes from liquid into vapor is encountered. Even though the basic concept is the same however there are many processes which utilize that concept in application today, the main thermal desalination technologies are Multi-Stage Flash Distillation (MSF), Multiple-Effect Distillation (MED), and Vapor Compression (VC).

4.2 Membrane desalination processes

With the growth of membrane science, RO overtook MSF as the leading desalination technology, membrane desalination processes in general and commercial RO processes in particular have been undergoing appreciable development. Important factors in the expansion of commercial RO applications are their favorably low power requirements and the realization of continuous technical improvements in membranes which are used in RO systems, RO was first applied to brackish groundwater with first large scale plants in late 1960s. A decade later RO membrane after further development was suitable for seawater desalination and become a strong competitive to conventional thermal desalination by the 1980s (Vandercaseele & Bruggen, 2002) and hence was able to expand the water sources used for desalination and utilize the brackish groundwater and dominate its market.

In the last two decades, RO processes have undergone major advancements significantly enabling now the massive production of water with a moderate cost, providing flexible solutions to different necessities within the fields of population supply, industry and agriculture. The great development of reverse osmosis (RO) technology has been a consequence of several factors such as energy consumption reduction, improvement of membrane material, and decrease in membrane cost.

Membrane processes are basic in principles, where semi-permeable membrane is used to allow the passage of water but not the salt to under certain driving forces. Different types of membrane processes include reverse osmosis RO, nanofiltration, and forward osmosis FO; the later is still under development mainly at laboratory and pilot scales.

4.2.1 Reverse osmosis RO membrane desalination

Osmosis is a natural physical process, where the solvent (i.e. water) moves through semi-permeable membrane (i.e. permeable to solvent and impermeable to solute) from low solute concentration to higher concentration creating differential pressure called osmotic pressure. The osmotic pressure depends mainly on the concentration difference, temperature, and nature of solute). The process continues till the hydraulic pressure difference due to the liquid column is equal to the osmotic pressure. In reverse osmosis RO, hydraulic pressure in value greater than the osmotic pressure is applied to the concentrated solution, results in reversing the osmosis process and in net solvent or water flow from the concentrate to the dilute solution (Fritzmann et al., 2007).

RO membranes do not have a distinct pore that traverse the membrane; it consists of polymeric material forming a layered, web-like structure characterized by high rejection to most of the dissolved solids present in water, with typical salt rejection above 99 % (Lee et al., 2011). The driving force for RO process is the applied hydraulic pressure which varies considerably from 15-25 bar for brackish water, and 60-80 bar for seawater (Fritzmann et al., 2007).

4.2.2 Nanofiltration NF membrane desalination

The Nanofiltration NF term was first introduced by FilmTech in 1980s to describe RO membranes that allow selectively and purposely some ionic solutes to path through the membrane, the membrane's selectivity was towards solute of about 1 nm cutoff, and hence the term nano comes from (Wang, 2008). Nanofiltration is an intermediate between RO membranes which has a low Molecular Weight Cut Off MWCO of about 100 and ultrafiltration membranes which has MWCO of about 1000 (Eriksson, 1988).

NF membranes have higher permeability for monovalent ions such as Na, K, and Cl and very low permeability to multivalent ions such as Ca, Mg, SO_4 and organics with MWCO of around 300 (Rautenbach & Groschl, 1990), as a result NF membranes were used mainly for removal of hardness and natural organic matter, as pretreatment before RO and MSD for seawater desalination (Al-shammiri et al., 2004; Hassan et al., 1998), and for groundwater quality enhancement (Burggen & Vandercasteele, 2003; Gorenflo et al., 2002; Saitua et al., 2011; Tahaikt et al., 2007). Nanofiltration membranes operating in a similar fashion to that of RO membranes, except of lower driving pressure and hence lower energy requirements, higher flowrates, and lower product water quality (Schaep et al., 1998).

4.2.3 Electrodialysis/ electrodialysis reversal membrane desalination

Electrodialysis/Electrodialysis Reversal membrane desalination (ED/EDR) is one of the oldest tried desalination processes, with ED process since the 1950s. In ED/EDR electrochemical separation is the main phenomena takes place utilizing electrical power as driving force to separate ions through ion-exchange membranes (Gleick, 2001). In typical ED cell, a series of anion- and cation-exchange membranes are a arranged in alternating pattern between the two electrical electrodes, anode and cathode, and hence ion concentrations increase in alternating compartments, and decrease simultaneously in the other compartments (Walha et al., 2007), ED process has been applied successfully but on small scale for brackish water (Adhikary et al., 1991; Brown, 1981; Harkare et al., 1982), and seawater (Sadrzadeh & Mohammadi, 2008; Seto et al., 1978).

In the 1970s, electrodialysis reversal EDR has been introduced as an innovative modification to the conventional electrodialysis, EDR operates on the same principle as ED. However the polarities of the electrodes are reversed for short time at specified time intervals, so that the ions are attracted in the opposite direction, and hence the brine and product channels are switched (Katz, 1979). EDR process has several features that promoted its application such as ability to treat feedwater of different qualities i.e. higher content of dissolved and suspended solids, ability to operate with high salts saturation levels and hence higher scale resistance, chlorine tolerance, cleanability, higher recovery, un affected by non-ionic species such as silica, low chemical pretreatment, and durability (Buros, 1999; Fubao, 1985; Katz, 1979; Katz, 1982; Valcour, 2010). In addition ED/EDR have been integrated successfully with other desalination processes such as RO (Oren et al., 2010).

4.2.4 Forward osmosis FO membrane desalination

Although RO membrane desalination has the major share in membrane desalination plants, the energy and membrane replacement cost are of major concern, and hence there is a search for new low energy and low fouling membrane processes. Forward osmosis (FO) or called direct osmosis (DO), employs the natural physical osmosis process by increasing the osmotic pressure in the permeate side to balance the pressure in the opposite side.

The FO exploits this natural tendency of water to move through the semi-permeable membrane from the saline water to a more concentrated solution called draw solution, the draw solution has significantly higher pressure than the saline water. Draw solutions of different natures have been tested for FO operation. Volatile solutes or gases such as sulfur dioxide which can be stripped out later to have pure water, perceptible salts such as aluminum sulfate which can be treated by lime to precipitate aluminum hydroxide and calcium sulfate, a two-stage system with sulfur dioxide-potassium nitrate used as the draw solutions has been also evaluated (McCutcheon et al., 2005a).

However recently more attention has been given to ammonia-carbon dioxide system due to several advantages such as the high solubility of ammonia and carbon dioxide gasses in water and the solution of formed ammonium bicarbonate have a high osmotic pressure which in turn provide higher water flux and recovery, followed by ease separation of the gases by moderate heating, which will be recycled back to the process (McCutcheon et al., 2005b; McGinnis & Elimelech, 2007).

Unlike other membrane process, FO does not require any hydraulic pressure to be applied and hence less energy requirements, which in turn results in less capital and operating costs. Furthermore the process has much lower fouling propensity (McGinnis & Elimelech, 2007, Phillip et al., 2007). In addition FO has been integrated successfully with other desalination processes such as RO enabling increased recovery, lower energy consumption (Lee et al., 2009; Martinetti et al., 2009; Tang & Ng, 2008; Yangali et al., 2011).

4.3 Other desalination processes

There are many other desalination process that are available today, some of them still at research and development stages, however they did not reach the development level to be commercialized on large scale as the previous processes, although many of them have a very promising features over the now widely used desalination process such as less corrosion and scaling problems, less energy consumption, less need for pre and post treatment. These processes include solar, ion-exchange, freezing, and membrane distillation.

5. Developments in desalination processes

Desalination has been extensively used over the past decades; with the great developments in desalination industry that have led to a higher acceptance and growth worldwide, particularly in arid areas. Although different desalination processes are well established today, further development are needed to resolve its various technical and operational issues which represent the essential key for successful desalination, including feed characterization for fouling and scaling propensity, process development, energy requirements, desalination economics, and finally brine disposal which will be given more attention due to its high importance (Sheikholesami, 2009).

5.1 Scaling and fouling

Fouling is a phenomenon that plagues the operation of desalination units, the deposition of foulants on to the heat (in case of thermal desalination) and mass (in case of membrane desalination) transfer surface results in reducing water productivity and decrease product quality, therefore as the fouling deposit builds up the energy consumption increases to accommodate for the required product flow till the unit is cleaned (Hamrouni & Dhahbi, 2001; Sheikholesami, 2004).

Fouling of membrane surface can be caused by any of the rejected constitutes, and can be generally classified to chemical fouling or scaling caused by sparingly soluble inorganic salts exceeding their saturation level, physical or colloidal fouling caused by particulate matter, biological fouling or biofouling due to the formation of biofilms of microorganisms, and finally organic fouling caused by natural organic matter NOM (Fritzmann et al., 2007). The water recovery is mainly constrained by fouling, and hence it is paramount to mitigate fouling of desalination units (Semiat et al., 2004).

The general approach to avoid scaling and fouling by sparingly soluble salts, is to estimate the saturation level of these salts according to the feed water quality, design recovery, and operation conditions and try to operate below such saturation levels where the solution is stable (Sheikholesami, 2004). Extensive work has been done to study the fouling and to determine the saturation levels of common scale forming sparingly soluble salts at different conditions with focus on calcium, barium, and silica particularly calcium sulfate, barium sulfate, calcium carbonate, and silica.

In seawater desalination main types of fouling is scaling by calcium carbonate, calcium sulfate, and magnesium hydroxide in case of thermal desalination. For membrane desalination, the major fouling of concern is biofouling (Al-Ahmad & Abdil Aleem, 1993). Fouling caused by precipitation of sparingly soluble salts is less likely to occur mainly due to the relative lower recovery, higher ionic strength, and low bicarbonate and sulfate concentration (Reverter et al., 2001). The contribution of each type of fouling in typical seawater RO desalination are 48% for biofouling, 18% for inorganic colloids, 15% for organic matter, 13% for silicates, and only about 6% for mineral deposits (Shon et al., 2009). Hence one of the most important steps in seawater pretreatment for desalination is disinfection with optimized dose of biocide, usually chlorine, in order to reduce biofouling propensity (Fujiwara & Matsuyama, 2008).

Brackish groundwater however has higher quality as it is mainly characterized by low content of suspended solids, low bacterial count, and low content of organic matter; as a result the most found type of fouling is scaling by sparingly soluble inorganic salts such as calcium and barium salts and silica.

The dissolved solids present in groundwater results mainly from chemical weathering or dissolution of geological formations i.e. minerals which can be attributed to the direct contact of groundwater with the calcium carbonate, and calcium sulfate rocks forming the aquifer. In addition sulfate may result from biological oxidation of reduced sulfur species. As a result the different aspects of scaling by calcium sulfate and calcium carbonate has been intensively studied (Sheikholesami, 2003a; Sheikholesami, 2003b; Rahardianato et al., 2008).

Silica originates from the dissolution or chemical weathering of amorphous or crystalline SiO_2 and the major clay minerals (Faust & Aly, 1998). Crystalline silica has a very low solubility in water, however amorphous silica can have solubility up to 120 mg/L at pH 7 and the solubility increases with pH increase reaching around 889 mg/L at pH 10 (Hamrouni & Dhabi, 2001; Sheikholesami & Tan, 1999). Silica in water can be classified into two categories 1) soluble or dissolved silica which contains monomers, dimmers, and polymers of silicic acid, and 2) insoluble or colloidal silica, which results from high polymerization of silicic acid. Due to its severe effect on membrane desalination performance and lifetime of membranes, great attention has been paid to silica fouling (Al-Shammiri et al., 2000; Ning, 2002; Semiat et al., 2003; Sheikholesami et al., 2001,).

In any desalination process, there are three main factors for sustainable operation: 1) proper design, 2) proper pretreatment, and 3) proper operation and maintenance; with the proper pretreatment as the foundation for successful operation (Neofotistou & Demadis, 2004). The primary goal of pretreatment is to lower the fouling propensity during the desalination process, and the required pretreatments depend mainly on the characteristics of the water resource (Greenlee et al., 2009). Scale inhibitors or antiscalents are chemicals that are added to water during pretreatment to prevent scale formation and usually work synergically with dispersant polymers.

However most of traditional antiscalents are successful in scale control for crystalline mineral precipitates but not silica because it is amorphous (Freeman & Majerle, 1995), and hence control of silica scaling requires chemical pretreatment.

5.2 Process development

Desalination is a multi unit process, starting from water intake, pretreatment, desalination, and post-treatment. Desalination plants in the past used to contain one type of desalination processes in the past. However attention has been recently paid to hybridization of different desalination processes together, with main objective of maximizing overall recovery, minimizing energy requirements, and cost reduction. NF has been integrated successfully with RO process mainly as pretreatment step, which resulted in improving the RO performance (Al-Shammiri et al., 2004; Hassan et al., 1998), EDR with RO (Oren et al., 2010), FO with RO (Lee et al., 2009; Martinetti et al., 2009; Tang & Ng, 2008, Yangali et al., 2011) in case of membrane desalination processes, integration of VC with MSF and MED (El-Dessouky et al., 2000; Mabrouk et al., 2007), and even combination of thermal and membrane processes by integrating NF/RO/MSF together (Hamed et al., 2009), and FO and MD to RO (Martinetti et al., 2009).

In membrane processes, development of much better membrane material that can work for wider range of pH, chlorine resistant, high mechanical strength to withstand higher hydraulic pressures, better salt rejection, and scale resistant are considered to be the next breakthrough in membrane desalination development, enabling higher recoveries at lower cost (Sheikholesami, 2009).

5.3 Energy requirements

Desalination processes are known for their intensive energy consumption especially thermal desalination, and hence energy consumption make up the major part of operation cost of any desalination process, the energy consumption differs according to desalination process in use i.e. thermal or membrane, water source and quality i.e. seawater or brackish water, design recovery, system design, plant capacity, and utilization of energy recovery devices.

Energy requirement for desalination processes is generally reported as specific energy consumptions in kWh/m^3 of product water. There are a wide range of reported values for energy consumption in desalination with the most recent values of about 1.8 kWh/m^3 for seawater desalination using MED, 4 kWh/m^3 for MSF with heat recovery mechanism incorporated (Khawaji et al., 2008). However for seawater using RO it went down from 20 kWh/m^3 in early 1970s to 1.6-2 kWh/m^3 recently, and below 1 kWh/m^3 for brackish water with energy recovery devices (Fritzmann et al., 2007; Khawaji et al., 2008).

In RO operation, the main energy consumption is mainly for the high pressure pump to provide hydraulic pressure in excess to the osmotic pressure, most of this pressure is retained by the concentrate stream flowing out of the RO unit. Energy recovery devices ERD have been developed mainly for RO operation to recover some of the energy retained in the concentrate before disposal. There are two main classes of ERD, class I which transfer hydraulic energy from the concentrate stream to the feed stream in one step with net energy transfer of more than 95%. Class II transfer hydraulic energy of the concentrate to centrifugal mechanical energy and then to hydraulic energy in the feed in two steps process (Greenlee et al., 2009).

Integration of desalination plants with power plants or as called cogeneration, which refers to the use of single energy source for multiple needs; mainly encountered with thermal desalination offers better energy utilization. For example, power plants use high pressure steam for power generation by means of turbines, the steam comes out at low pressure which is very suitable for thermal desalination (Gleick, 2001). In addition there are efforts for implementing cogeneration in RO desalination process in order to address the water-electricity demand trade off (Altmann, 1997).

Use of renewable energy sources for driving the desalination process provide another development opportunity, specifically for membrane desalination where less energy is required and for rural communities in which desalination systems are generally small in size with usually non-continuous operation, and hence can be integrated to renewable energy sources. Considerable efforts has been made for integrating desalination processes with different renewable energy sources namely solar (photovoltaic and thermal), wind, and geothermal. The renewable energy can be used in one of two forms thermal or electrical, depending on which one that best match the desalination process (Al-karaghouli et al., 2009; Al-karaghouli et al., 2010; Forstmeier et al., 2007; Mathioulakis et al, 2007).

5.4 Desalination economics

Economics of any process represent the most crucial part for development and application, and hence economical feasibility is a very important factor when considering desalination processes. In desalination processes it is difficult to standardize the economics of process because it is case specific. There are several factors that affect desalination cost such as water source (brackish or seawater), desalination process used (thermal or membrane), energy source (traditional or renewable), and plant size (Dore, 2005; Karagiannis & Soldatos, 2008).

The source, and hence the quality of the feed water plays important rule for determination of both capital and operating cost, and the overall desalination cost. Brackish water has much low salt content and better water quality than seawater and therefore, it incorporates less capital and operating costs. The most recent data for average investment cost for brackish water was around $200-450/ (m^3/d) with product water cost of $0.25-0.75/m^3 for RO process which is the process dominating the brackish water desalination market (Vince et al., 2008; Yun et al, 2006).

The product water desalination cost varies significantly according to the salinity of the water sources. For example the product water cost for brackish water with salinity around 3,000 mg/L was found to be $0.32/m^3. However, for water with salinity around 10,000 mg/L, the desalination cost was $0.54 /m^3 (Karagiannis & Soldatos, 2008). The same trend was observed also for sea water desalination with desalination cost ranging from $0.54 /m^3 for Mediterranean seawater to $0.87/m^3 for Arabic Gulf seawater (Greenlee et al., 2009).

Capacity or size of desalination plant greatly affects the product water cost; table 3 shows the average desalination cost for brackish and seawater desalination plants of different production capacities.

Feed water	Plant size (m^3/ d)	Cost ($ / m^3)
Brackish	≤ 1,000	0.78 – 1.33
	5,000 – 60,000	0.26 – 0.54
Seawater	< 1,000	2.2 – 11.25
	1,000 – 5,000	0.7 – 3.9
	12,000 – 60,000	0.44 – 1.62
	> 60,000	0.50 – 1.0

Table 3. Size of desalination plant and water production cost (Karagiannis & Soldatos, 2008).

6. Brine disposal from inland desalination

Brine, concentrate, or reject are different names for a stream that is commonly produced from any desalination process. In any desalination process, two streams are produced: 1) product water with high quality, 2) brine or concentrate stream that contains all the salts were originally present in the feed water in addition to the chemicals added in the pretreatment and during desalination such as antiscalents.

To reduce energy consumption, cleaning time and expenditure, loss of production during downtime, it is paramount to mitigate fouling. The general approach is to study the feed water characteristics and couple it with the expected recovery and operating conditions; and to operate at conditions where the solution is stable, hence scaling and fouling is minimized. However working at lower recoveries to avoid the fouling of membrane will increase the flow of brine stream generated, which present the main trade-off for desalination operation.

Brine stream does not contains only 2 or 4-5 folds the salinity of the feed water as in case of seawater, or brackish water respectively, but it contains all the chemicals that has been added to the desalination process during pretreatment. Moreover in case of thermal desalination it will be at high temperature, and hence more attention should be taken when considering the brine discharge method (Ahmed et al., 2002).

The problem of brine discharge is different in sea water and brackish water desalination. In the case of seawater desalination plants the problem is readily solved since these plants are usually placed near the coast, so the discharge method of choose is usually to discharge it back to the sea through brine pipes or submarine emissaries. Encouraging facts to utilize that option are, first the discharged brine is of similar chemistry, even being more concentrated but only by 50-100 %. Second is that the volume of brine stream relative to the water body being discharged to being very small, hence lower drawbacks are expected. However, there are many criteria to be considered such as having the discharge point far enough from intake at good mixing zone so it can be mixed with the main body of seawater.

However, the management of brine from brackish desalination plants i.e. inland desalination can be significant problem in case they are placed far from the coast (inland plants). Some of the conventional options for brine disposal from inland desalination plants are: 1) disposal into surface water bodies, 2) disposal to municipal sewers, 3) evaporation ponds, 4) deep well injection, and 5) irrigation of plants tolerant to high salinities (Ahuja & Howe, 2005). The main factors that influence the selection of a disposal method, among others, are: 1) volume or quantity of concentrate, 2) quality and constitutes present in the concentrate, 3) physical and geographical considerations, 4) capital and operational costs, 5) availability of receiving site, 6) permissibility of the option, facility future expansion plan, and 7) public acceptance. All of these factors together will affect the cost of brine disposal that can ranges from 5 -33 % of the total desalination cost (Ahmed et al, 2001).

Brine disposal method should be considered after the necessary studies and investigations have been performed in order to minimize the brine stream to be disposed off, and hence reduce the cost of subsequent disposal. This is mainly achieved by employing the proper feed water pretreatment, proper desalination process, maximizing the system recovery. However attention to the increased salinity and quality of brine should be considered.

6.1 Disposal to surface water bodies and sewers systems

Disposal of brine to surface water bodies if a available, present the first option to choose as it represent a ready and good solution to the challenge of brine disposal taking into account that the brine stream is diluted by mixing with the water body. However many consideration should be taken into account. The salinity if the receiving body might increase due to the disposal of the high salinity brine, and hence the self-purification capacity of the receiving water should be considered (Ahmed et al., 2000). As a result disposal to surface water should be permitted only if that will avoid dramatic impact on environment.

Another option is to dispose the brine to the local sewage system, which is usually employed by small membrane desalination plants. This option has many advantages such as use of the ready available and installed sewage system, lowering the BOD of the domestic sewage water. However that should be practiced carefully as the salinity of sewage water might increase which might affect the wastewater treatment facility especially biological treatment step. This might also render treated municipal sewage effluent unsuitable for agriculture use when disposing brines with high salinity (Ahmed et al., 2000).

6.2 Disposal to evaporation ponds

In evaporation bonds, the brine is discharged into a large surface area pond, where the water is naturally evaporated. Use of evaporation pond technology is practiced primarily in the arid and semi-arid areas, particularly in Middle East and Australia. Evaporation pond is probably the most widespread method for brine disposal from inland desalination plants.

Simple evaporation ponds have many advantages such as being easy to construct, low maintenance and operation cost, no equipment are needed specifically mechanical. Making it the most appropriate method with lower cost, especially in arid areas with high evaporation rates, low rainfall, and low land cost (Ahmed et al., 2000). Use of evaporation ponds for cultivation of brine shrimps has been studies as well, giving ideal place for brine-shrimp production as it present mono-culture environment under natural conditions with absence of any food competitors or predators (Ahmed et al., 2001).

The basic concern associated with use of evaporation pond for brine disposal is leakage of brine through soil. This may result in subsequent contamination and increasing salinity of the aquifer. Electrical conductivity and concentration of salts in the evaporation ponds can be used as indicators for leakage in the pond, where insignificant increase is a strong indication of brine leakage through the soil (Amed et al., 2001). Deterioration of soil and groundwater quality in areas nearby evaporation ponds used for brine disposal in KSA, UAE, and Oman was investigated and reported as one of the draw backs to use of evaporation bonds (Al-Faifi et al, 2010; Mohamed et al., 2005).

As a result most of the evaporation ponds installed recently are lined with polymeric sheets. Liner installation should be carried out carefully as joints sealing is very important for leakage prevention. Furthermore double lining is strongly recommended with proper monitoring for leakage.

In addition reduction in production from agricultural lands caused by deposition of airborne salts from dried concentrate of evaporation bonds, and formation of eyesores caused by improper disposal of concentrates on nearby land can be another disadvantage of brine disposal to evaporation ponds.

In conclusion, while operation and maintenance costs for evaporation ponds are minimal, large land areas are required which increases as the plant capacity increase, and pond construction costs are high due to lining and monitoring requirements. Consequently, proper evaporation ponds are often cost prohibitive and impractical for handling significant concentrate flow. Furthermore, water evaporated from a pond is often a lost resource.

6.3 Deep well injection

In deep well injection, the brine is injected back underground to depth ranges from few hundreds of meters to thousands of meters, depending on many factors which should be considered while designing, installing, and operating the system. Deep well injection for brine disposal includes permitting considerations, which look for identification of adequate geologic confining unit to prevent upward migration of effluent from the injection area. While design considerations focus generally on the tubing and packing installed inside the final cemented casing of the injection well, compatibility of the concentrate with the tubing material (to avoid corrosion), expected concentrate flow, and leak detection and monitoring systems (Skehan & Kwiatkowski, 2000).

One of the very attractive options with deep well injection is to use depleted oil and gas fields for brine disposal. This encounter many advantages such as making use of the readily available gas and oil wells, long experience encountered with the operation of such wells. However before applying this option the fields should be tested physically and chemically for accepting the brine stream [Mace et al., 2006; Nicot & Cjowdhury, 2005]

Generally site selection for installing of such deep well, is the most important step, and hence hydrological and geological conditions should be considered, as example the wells should never be installed in areas vulnerable to earthquakes (Ahmed et al., 2000). Although of availability of such option to many inland desalination plants, however many factors should be considered with deep well injection for brine disposal which can be summarized as follow (Mickley et al, 2006):

1. Site selection, which is performed through many geological and hydrological studies, to identify the proper area for installing the well,
2. High cost, associated with both capital and operational cost,
3. Possibility of corrosion and subsequent leakage in the well casing,
4. Seismic activity which could cause damage to the well and subsequently result in leakage,
5. Uncertainty of the well life,
6. Pollution of groundwater resources, which may result from high salinity and the presence of other harmful chemicals in the brine.

6.4 Land applications of brine

Land application such as use in irrigation systems that was originally developed for sewage effluents, can be used for brine disposal, and hence helps conserve natural resources. In areas where water conservation is of great importance, spray irrigation is especially attractive option. Concentrate can be applied to cropland or vegetation by sprinkling or surface techniques for water conservation when lawns, parks, or golf courses are irrigated and for preservation and enlargement of greenbelts and open spaces. Crops such as water-tolerant grasses with low potential for economic return but with high salinity tolerance are generally chosen for this type. However soil sanlinization and groundwater contamination should be carefully considered (Mickley et al, 2006).

7. Inland desalination with zero liquid discharge

In many cases of brackish water desalination, brine management is critical and of high concern, and hence the need for affordable inland desalination has become critical in many regions of the world where communities strive to meet rapidly growing water demands with limited freshwater resources.

Where brine disposal and management is a problem, given the disadvantages of existing brine disposal and management methods, it is imperative to find alternative Zero Liquid Discharge ZLD technologies that provide more affordable concentrate management. In ZLD, brine is treated to produce desalinated water and essentially dry salts; therefore there is no discharge of liquid waste from the site. Most ZLD applications in operation today treat industrial wastewater using thermal or membrane separation processes, or a combination of these technologies.

Thermal desalination is a mature technology that has been practiced for long time especially where energy is relatively inexpensive, while it is a proven process that generates high quality product water, thermal desalination is energy-intensive and its capital and operating costs are high. Membrane processes has been proved to provide high quality water, but also has some limitation concerning scaling and maximum hydraulic pressure and cannot alone provide ZLD solution. Advancement of ZLD science and associated reduction of ZLD costs will be of tremendous benefit and will alleviate the water supply challenges faced by many communities worldwide.

ZLD desalination present the perfect solution for the brine disposal and management problem usually encountered with inland desalination plants. In addition applying inland desalination with ZLD provide several advantages, the main advantages can summarized as below:

- Maximize Water Recovery: with ZLD systems approaches 100 % recovery, when compared to the conventional Inland desalination system with regular recovery of about 70-85%. ZLD systems should be able to provide more product water or less plant size by 15-30 %.
- Preserving Natural Resources: for inland desalination with ZLD systems, the natural resources, which are mainly groundwater and land, are preserved both quantitatively and qualitatively, by avoiding the different problems associated with conventional brine disposal methods.
- Byproduct salts: the ZLD system results into two stream, product water, and dry salts, these salts can be treated as added value product rather than solid waste, finding a lot of applications and beneficial uses.
- Integerability and applicability: ZLD system can be integrated to any existing inland desalination plant of any size and location. This is mainly because the system operates

on treat the brine resulted from the existing desalination plants, and hence it can be integrated at any stage from design to operation stages.

There have been many attempts to achieve a successful inland desalination with ZLD, however more attention and further research work and process developments are needed in order develop a full economic-technical feasible ZLD desalination. In the following sections the current efforts for providing a ZLD system, as well as further developments and research needs will be discussed.

7.1 Current zero liquid discharge schemes for inland desalination

Little literature work is available on ZLD systems for inland desalination; however three main schemes can be concluded and summarized as follow:

- Applying thermal processes directly to the brine generated from the primary desalination process, usually RO, followed by thermal processes for brine concentration, then crystallization or drying for final salt production (Mickley et al, 2006).
- Applying chemical treatment to the brine stream, followed by further membrane desalination, brine concentration, and finally crystallization or drying for final salt production (Bond & Veerapaneni, 2008; Mohammadesmaeili et al., 2010).
- Applying ED/EDR process to the brine stream making use of higher recovery encountered with such units, followed by crystallization or drying for final salt production (Greenlee et al., 2009, Oren et al., 2010).

Similarities between schemes are clear, especially for brine concentration and final salt production, with the difference mainly in brine treatment and further desalination. However brine treatment and further desalination results in significant reduction in the volume of brine to undergo the brine concentration and final drying/crystallization. Options for integrating different units in different setups can be investigated with an overall objective of ZLD desalination and production of salts can be generated, and should be evaluated for process optimization (Kim, 2011).

Conventional inland desalination system usually achieve 70-85 % recovery of the feed water, which is the largest recovery increment in single step, the recovery mainly depend on the quality of the feed water. However this recovery is usually limited due to scaling by sparingly soluble salts, typically calcium salts such as calcium sulfate and carbonate, in addition to silica (Freeman & Majerle, 1995; Rhardianato et al., 2008; Sheikholesami, 2003a; Sheikholesami, 2004). With this recovery range, about 15-30 % of the feed stream will be rejected as brine which should be disposed off.

In the first ZLD scheme this 15-30 % is fed to thermal processes using single or multiple effect evaporators or vapor compression evaporators for brine concentration which to be followed by crystallization or drying to obtain final dry salts. In the second scheme the brine is treated chemically to remove most of scale forming constitutes, achieving high removal of such constitutes rendering the treated brine suitable for further membrane separation to recover more water. With the two membrane desalination process with the intermediate brine treatment step recovery up to 95% can be achieved, moreover making use of membrane desalination reduces the cost, and minimizes the energy requirements. Moreover reduces the volume of brine to be handled by final evaporation step which results in lowering the energy requirements and hence the overall process cost. In the last scheme ED/EDR unit are employed, which can operate at high saturation levels of sparingly soluble salts as in the case of brine streams, and where high recovery up to 97% can be achieved.

7.2 Precipitation softening for brine treatment in zero liquid discharge systems

Intermediate brine treatment step as employed in the second ZLD scheme is the one receiving large attention recently. The main objective of this step to remove most of the scale forming constitutes typically calcium, magnesium, carbonate, sulfate, and silica. It is hard to find a chemical treatment process that is able to efficiently remove all of these constitutes. Furthermore most of the tested chemical treatment processes were not able to completely remove such constitutes. However the achieved removal efficiency was good enough to prevent such constitutes from limiting the recovery in the secondary membrane desalination process.

Precipitation softening is one of the widely used processes for reduction of hardness (calcium and magnesium) and alkalinity (mainly bicarbonate) in water treatment plants. The reduction of hardness is mainly achieved by removal of calcium as calcium carbonate $CaCO_3$ and magnesium as magnesium hydroxide $Mg(OH)_2$. This is usually achieved by addition of alkali usually lime, calcium hydroxide $Ca(OH)_2$ in lime softening or sodium hydroxide, NaOH in caustic softening and sodium carbonate Na_2CO_3 depending on the quality of water to be treated (Reynolds & Richards, 1996). Removal of calcium and magnesium present one of the major targets for brine chemical treatment, particularly calcium, as magnesium cause scaling problems only at high pH values forming insoluble magnesium hydroxide $Mg(OH)_2$. However at the normal pH values found in brine streams it will be mainly saturated by calcium sulfate and carbonate (Sheikholesami, 2003a; Sheikholesami, 2004; Rharadianato et al., 2008).

Silica removal during the precipitation softening was extensively studied, and it was found that silica is removal could be by co-precipitation with metal hydroxides, specifically iron, manganese, and magnesium hydroxides, or could be by precipitation as magnesium and calcium silicate (Sheikholesami & Bright, 2002). Furthermore caustic softening using only sodium hydroxide was found to be more effective and more viable in removal of silica over lime-soda softening using lime and soda ash (Al-Rehaili, 2003; Sheikholesami & Bright, 2003). Addition of sodium aluminate and aluminum sulfate was found to enhance the removal of silica during the softening process by co-precipitation with aluminum hydroxide (Cheng et al., 2009; Lindsay & Ryznar, 1939). Conventional softening process is slow, requires extensive space, and generates large volume of sludge which will need dewatering and further treatment later on (Kadem & Zalmon, 1997). As a result a more advanced process designated Compact Accelerated Precipitation Softening CAPS was developed to enhance the performance of the precipitation softening process. In CAPS process the saturated solution is passed through cake of calcium carbonate to enhance crystallization and approach equilibrium rapidly, the process has been found to overcome the different disadvantages encountered in conventional softening process (Gilron et al., 2005; Masarawa et al., 1997; Oren et al., 2001).

Although different precipitation softening processes have been applied basically for surface water treatment and as a pretreatment for membrane processes, specifically NF and RO (Al-

Rehaili, 2003; Cheng et al, 2009; Gilron et al., 2005; Kadem & Zalmon, 1997; Masarawa et al., 1997; Oren et al., 2001; Sheikholeslami & Bright, 2002) showing high efficiency in removal of calcium, magnesium, silica, and heavy metals. However such softening processes were found to be very effective and promising when applied for brine treatment where high calcium, magnesium, and silica removals from brine streams has been achived enabling higher recovery in the subsequent membrane desalination, facilitate reaching zero liquid discharge desalination [Comstock et al., 2011; Gabelich et al., 2007; Ning et al., 2006; Ning & Tryoer, 2009; Rahardianto et al., 2007].

Sulfate usually present in the brine streams in high concentrations, relative to those of calcium, magnesium, carbonate, and silica. However in presence of calcium, saturation and hence scaling due to calcium sulfate is very likely to happen (Rahardianato et al., 2008; Sheikholesami, 2003a; Sheikholesami, 2004). Precipitative softening processes were found to be very effective in removal of calcium, magnesium, carbonate, and silica. However such process had no success for removal of sulfate, even though removal of calcium from brine stream reduces the scaling potential of calcium sulfate. However it will be paramount to remove sulfate completely or partially, converting the brine chemistry typically to monovalent ions i.e. sodium, potassium, and chloride which has no scaling potential at the normal membrane desalination operating conditions.

Several works has been performed on removal of sulfate from industrial wastewater streams such as paper mills, mining, and fertilizers, and several attempts have been performed to reach zero discharge with such streams using membrane and thermal separation processes (Ericsson & Hallmans, 1996). However many attempts has been worked to employ precipitation and crystallization removal of sulfate as calcium sulfate, gypsum, by addition of calcium mainly as calcium hydroxide, lime (Tait et al., 2009), or as calcium chloride (Benatti et al., 2009), which was found to be very effective in removal of sulfate. Removal of sulfate as calcium sulfate below 1300 mg/L was found to be very hard due to solubility limits. However addition of aluminum as aluminum sulfate or alum, aluminum chloride, aluminum nitrate (Christoe, 1976), and sodium aluminate (Batchelor et al, 1985) was found to enhance the removal of sulfate far below this value by formation of more complex solids (Batchelor et al, 1985; Christoe, 1976) which has much lower solubility compared to that of calcium sulfate precipitated by lime addition only.

7.3 Secondary brine concentration and final salt production

The treated brine after being further concentrated in secondary membrane desalination has to be further concentrated reaching zero liquid. Zero liquid and dry salts cannot be produced by membrane desalination such as RO or NF, and hence the concentrated brine has to be subjected to thermal process such as brine concentration followed by crystallization or drying to produce dry salts.

Thermal processes such as single and multi stage evaporators, or vapor compression evaporators are usually employed for further brine concentration. Such units have dual purpose objectives which are further recovery of water with very high quality with salinity about 10 mg/L, and brine concentration up to 250,000 mg/L, with recovery above 90%. Final salt production can be achieved after brine concentration which usually performed in crystallizers or dryers (Mickley, 2006).

In addition to the salts produced from final crystallizer/dryer that can be assumed as byproducts from ZLD desalination, the precipitated solids from brine chemical treatment can be considered as another byproduct or added value product. This precipitate is rich in calcium as calcium carbonate, magnesium as magnesium hydroxide, and silicate of calcium and magnesium, in addition to gypsum or calcium sulfoaluminate in case of sulfate removal, such mixture can find a wide range of applications such as road pavement, cement industry, and any other applications where there is a need for mixture of similar composition.

7.4 Cost associated with zero liquid discharge systems

Reaching inland desalination with zero liquid discharge has to be considered on both scales, technically and economically, while technical ZLD system can be successfully achieved through the different ZLD schemes. However the costs associated with each proposed ZLD system should be carefully considered. It easily noticeable that employing a secondary membrane desalination step is of high importance for reduction of both capital and operating costs over the conventional thermal ZLD systems due to the reduced volume of brine stream to be thermally treated.

A cost comparison for standard bench mark brine treatment by brine concentration and evaporation to advanced brine treatment using secondary RO desalination and final brine concentration and drying for brine of different qualities has been performed. The study showed that a cost reduction ranging from 48-67%, with reduction in energy requirement of 58 - 72% using the advanced ZLD system (Bond & Veerapaneni, 2008) can be achieved. However it worth to mention that the comparison was for the brine management only, not the whole inland desalination system, as the primary RO desalination is a kind of standard step employed for all ZLD schemes.

Inland desalination with zero liquid discharge usually has higher product water cost when compared to conventional inland desalination systems with no brine disposal is employed, but becomes very economically attractive when compared to the different brine disposal methods. The high cost mainly due to the fact that several units such as chemical treatment, secondary membrane desalination, brine concentration, and crystallization/drying are employed to recover only 15-30%. Which increase both capital and operating cost increasing the average product cost compared to single step desalination unit recovering 70-85 % (Greenlee et al., 2009). However due to the different strict regulation on brine disposal using the conventional methods, and the efforts for preserving the groundwater resources, more driving force for advancement of ZLD systems are encouraged (Mickley, 2006).

7.5 Developments and research needs for zero liquid discharge desalination

Desalination with zero liquid discharge is the ultimate achievement for any inland desalination process. This will help to overcome the brine disposal limitations currently faced for applying inland desalination. Although different zero liquid discharge schemes are currently developed or under development, however further development are needed to resolve its various technical, operational, and economical issues. The essential key for successful ZLD inland desalination are brine treatment, process development, energy requirements, and process economics, which should be given more attention and further research and development efforts.

Precipitative softening processes have been widely used for treatment of primary brine stream, however softening process improvements through chemical doses optimization, testing different chemical reagents aiming at high efficiency in removal of scale forming constitutes should help improving the overall process performance. Furthermore other chemical treatment processes should be investigated which can result in better performance.

ZLD system usually employs different units with different nature, such as membrane and thermal process, liquid and solids handling. Process development should look at the different viable and optimum units arrangement and operation conditions with the objective reducing energy requirements and cost.

Thermal processes are usually employed in ZLD systems for further brine concentration up to level that can be handled by crystallizer or dryer. Such processes are known to be energy extensive, and hence reduction in energy requirements and utilization of renewable energy should help in reducing overall energy requirements.

Economics of ZLD process is very important factor in employing the ZLD for inland desalination. Reaching competitive overall cost for inland desalination with ZLD to that of conventional desalination should help in wider application of the process for inland desalination systems.

8. Conclusions

In conclusion, as groundwater presents the main source of potable water to communities that do not have access to surface water, the deterioration of groundwater quality, specifically salinity is of high concern, which leads to the use of desalination techniques to overcome such problem. The use of membrane desalination systems in general and reverse osmosis in particular is very beneficial due to capacity flexibility, lower energy requirements, and in turn lower cost for brackish groundwater desalination. However the generation of brine stream is the main problem facing such systems, and which should be managed properly, there are different ways for brine disposal. However each one has certain advantages and disadvantages that are a matter of question. Approaching inland desalination with zero liquid discharge presents the solution for having a perfect inland desalination system. Given such need it is imperative to find a zero liquid discharge treatment technologies that provide more affordable concentrate management at reasonable cost, and hence a very active area of research is going on to provide such solution.

9. References

Adhikary, S. K.; Narayanan, P. K.; Thampy,S. K.; Dave, N. J.; Chauhan, D. K.; & Indusekhar, V. K. (1991). Desalination of Brackish Water of Higher salinity by Electrodialysis. *Desalination*, Vol. 84, No. 1-3, (October 1991), pp. 189-200, ISSN: 0011-9164.

Ahmed, M.; Shayya, W. H.; Hoey, D.; Mahendran, A.; Morris, R.; Al-Handaly, J. (2000). Use of Evaporation Ponds for Brine Disposal in Desalination Plants, *Desalination*, Vol. 130, No. 2, (November 2000), pp. 155-168, ISSN: 0011-9164.

Ahmed, M.; Shayya, W. H.; Hoey, D.; Al-Handaly, J. (2001) Brine Disposal from Reverse Osmosis Desalination Plants in Oman and the United Arab Emirates, *Desalination*, Vol. 133, No. 2, (March 2001),pp. 135-147, ISSN: 0011-9164.

Ahmed, M.; Shayya, W. H.; Hoey, D.; Al-Handaly, J. (2002) Brine Disposal from Inland Desalination Plants, *Water International*, Vol. 27, No. 2, (June 2002), pp. 194 -201.

Ahuja, N.; Howe, K. (2005). Strategies for Concentrate Management from Inland Desalination, *Proceeding of Membrane Technology Conference & Exposition 2005"*, pp. 761-776, Phoenix, Arizona, USA, March 6-9, 2005.

Altmann and Water Cogeneration, T. (1997) A New Power Concept with the Application of Reverse osmosis Desalination, *Desalination*, Vol. 114, No. 2 , (December 1997), pp.139-144, ISSN: 0011-9164.

Al-Ahmad, M.; Abdil Aleem, F. (1993) Scale Formation and Fouling Problems Effect on the Performance of MSF and RO Desalination Plants in Saudi Arabia, *Desalination*, Vol. 93, No. 1-3, (August 1993), pp. 287-310, ISSN: 0011-9164.

Al-Faifi, H.; et al. (2010) Soil Deterioration as Influenced by Land Disposal of Reject Brine from Salbukh Water Desalination Plant at Riyadh, Saudi Arabia, *Desalination*, Vol. 250, No. 2, (January 2010), pp. 479-484, ISSN: 0011-9164.

Al-Karaghouli, A.; Renne, D.; Kazmerski, L. L. (2009) Solar and Wind Opportunities for Water Desalination in the Arab Regions, *Renewable and Sustainable Energy Reviews*, Vol. 13, No. 9, (December 2009), pp. 2397-2407, ISSN: 1364-0321.

Al-Karaghouli, A.; Renne, D.; Kazmerski, L. L. (2010) Technical and Economic Assessment of Photovoltaic-Driven Desalination Systems, *Renewable Energy*, Vol. 35, No. 2, (February 2010), pp. 323-328, ISSN: 0960-1481.

Al-Rehaili, A. M. (2003) Comparative Chemical Clarification for Silica Removal from RO Groundwater Feed, *Desalination*, Vol. 159, No. 1, (September 2003), pp. 21-31, ISSN: 0011-9164.

Al-Shammiri, M.; Safar, M.; Al-Dawas, M. (2000) Evaluation of Two Different Antiscalents in Real Operation at the Doha Research Plant, *Desalination*, Vol. 128, No. 1, (March 2000), pp. 1-16, ISSN: 0011-9164.

Al-Shammiri, M.; Ahmed, M.; Al-Rageeb, M. (2004) Nanofilteration and Calcium Sulfate limitation for Top Brine Temperature in Gulf Desalination Plants, *Desalination*, Vol. 167, No. 2, pp. 335-346, ISSN: 0011-9164.

Amer, K. M.; Al-Muraikhi, A.; Rashid, N. (2008) Management of Coastal Aquifers- the Case of a Peninsula- State of Qatar; *20th Salt water Intrusion Meeting*, Florida, USA, June 23-27, 2008.

Ayers, R.S.; Westcot, D.W., (1994), Water Quality for Irrigation 29 Rev.1, accessed 15.08.11, available from:
http://www.fao.org/DOCREP/003/T0234E/T0234E00.htm#TOC

Bart Weiss, P.G. (2002). Brackish Groundwater Desalination Water Supply, *Proceedings of AMTA 2002 Biennial Conference and Exposition "Water Quality Enhancement through Membrane Technology"*, pp. 0-44, Tampa, Florida, August 6-9, 2002.

Batchelor, B.; McDevitt, M.; Chan, D. (1985) Removal of Sulfate from Recycled Cooling Water by the Ultra-High Lime Process, *Proceedings water Reuse Symposium III, AWWA Research Foundation,* pp. 798-812, Denver, Colorado, USA, 1985.

Benatti, C. T.; Tavares, C. R. G.; Lenzi, E. (2009) Sulfate Removal from Waste Chemicals by Precipitation, *Journal of Environmental Management,* Vol. 90, No. 1, (January 2009), pp. 504-511, ISSN: 0301-4797.

Bond, R.; Veerapaneni, S. (2008) Zeroing in on ZLD Technologies for Inland Desalination, *American Water Works Association Journal,* Vol. 100, No. 9, (September 2008), pp. 76-89, ISSN: 1551-8833.

Brich, R. P.; Al-Arrayedh M.; Hallmans B. (1985) Bahrain's Fresh Groundwater Situation and the Investigations into Using Bahrain's Brackish water Resources as a Feedwater for the Reverse Osmosis Desalination Programme, *Desalination,* Vol. 55, No. 1, pp. 397-427, ISSN: 0011-9164.

Brown, D. R. (1981) Treating Bahrain Zone C Groundwater using the EDR process, *Desalination,* Vol. 38, (November 1981), pp. 537-547, ISSN: 0011-9164.

Bruggen; B., Vandecasteele, C. (2003) Removal of Pollutants from Surface Water and Groundwater by Nanofiltration: Overview of Possible Applications in the Drinking water Industry, *Environmental Pollution,* Vol. 122, No. 3, (April 2003), pp. 435-445, ISSN: 0269-7491.

Buros, O.K. (1999) The ABCs of Desalting, International Desalination Association, accessed 15.08.11, available from: http://www.idadesal.org/pdf/ABCs1.pdf

Cheng, H.; Chen, S.; Yang, S. (2009) In-Line Coagulation/Ultrafiltration for Silica Removal from Brackish Water as RO pretreatment, *Separation and Purification Technology,* Vol. 70, No. 13, (November 2009), pp. 112-117, ISSN: 1383-5866.

Christoe, J. R. (1976) Removal of Sulfate from Industrial Wastewaters, *Journal of Water Pollution Control Federation,* Vol. 48, No. 12, (December 1976), pp. 2804-2804, ISSN: 0043-1303.

Comstock, S. E. H.; Boyer, T. H.; Graf, K. C. (2011) Treatment of Nanofiltration and Reverse Osmosis Concentrates: Comparison of Precipitative Softening, Coagulation, and anion Exchange, *Water Research,* Vol. 45, No. 16, (October 2011), pp. 4855-4865, ISSN: 0043-1354.

Delleur, J. W. (2nd Ed.). (2007). *The handbook of Groundwater Engineering,* CRC Press, ISBN 0-8493-4316-x, Florida, USA.

Dore, M. H. I. (2005) Forecasting the Economic Costs of Desalination Technology, *Desalination,* Vol. 172, No. 3, (February 2005), pp. 207-214, ISSN: 0011-9164.

El-Dessouky, H. T.; Ettouny, H. M.; Al-Juwayhel, F. (2000) Multiple Effect Evaporation – Vapor Compression Desalination Processes, *Chemical Engineering Research and Design,* Vol. 78, No. 4, (May 2000), pp. 662-676, ISSN: 0263-8762.

Eriksson, P. (1988) Nanofiltration Extends the Range of Membrane Filtration, *Environmental Progress.* Vol. 7, No. 1, (February 1988), pp. 58-62, ISSN: 0278-4491.

Ericsson, B.; Hallmans, B. (1996) Treatment of Saline Wastewater for Zero Discharge at the Debiensko Coal Mines in Poland, *Desalination,* Vol. 105, No. 1-2, (June 1996), pp. 115-123, ISSN: 0011-9164.

Faust, S. D.; Aly, O. M. (1998), *Chemistry of Water Treatment* 2nd Ed., Lewis Publishers CRC Press LLC, ISBN 1-57504-011-5, Florida, USA.

Food and Agriculture Organization of the United Nations FAO, AQUASTAT Program, (n.d.). Accessed 13.08.11, available from: http://www.fao.org/nr/water/aquastat/main/index.stm

Forstmeier, M; et al. (2007) Feasibility Study on Wind–Powered Desalination, *Desalination*, Vol. 203, No. 1-3, (February 2007), pp. 463-470, ISSN: 0011-9164.

Freeman, S. D.; Majerle, R. J. (1995) Silica Fouling Revisited, *Desalination*, Vol. 103, No. 1-2, (November 1995), pp. 113-115, ISSN: 0011-9164.

Fritzmann, C.; Lowenberg, J., Wintgens, T; Melin, T. (2007) State-of-the-Art of Reverse Osmosis Desalination, *Desalination*, Vol. 216, No. 1-3, (October 2007), pp. 1-76, ISSN: 0011-9164.

Fubao , Y. (1985) Study on Electrodialysis reversal (EDR) Process, *Desalination*, Vol. 56, pp. 315-324, ISSN: 0011-9164.

Fujiwara, N.; Matsuyama, H. (2008) Elimination of Biological Fouling in Seawater Reverse Osmosis Desalination Plants, *Desalination*, Vol. 227, No. 1-3, (July 2008), pp. 295-305, ISSN: 0011-9164.

Gabbrielli, E. (1981) A Tailored Process for Remineralization and Potabilization of Salinated Water, *Desalination*, Vol. 39, (December 1981), pp. 503-520, ISSN: 0011-9164.

Gabelich, C. J.; Williams, M. D.; Rahardianto, A.; Franklin, J. C.; Cohen, Y. (2007) High Recovery Reverse Osmosis Desalting Using Intermediate Chemical Demineralization, *Journal of Membrane Science*, Vol. 301, No. 1-2, (September 2007), pp. 131-141, ISSN: 0376-7388.

Gabelich, C. J.; Rahardianto, A.; Northrup, C. R.; Yun, T. I.; Cohen, Y. (2011) Process Evaluation of Intermediate Chemical Demineralization for Water Recovery Enhancement in Production-Scale Brackish Water Desalting, *Desalination*, Vol. 272, No. 1-3, (May 2011), pp. 36-45, ISSN: 0011-9164.

Gilron, J.; Daltrophe, N.; Waissman, M.; Oren, Y. (2005) Comparison Between Compact Accelerated Precipitation Softening (CAPS) and Conventional Pretreatment in Operation of Brackish Water Reverse Osmosis (BWRO), *Industrial Engineering and Chemistry Research*, Vol. 44, No. 15, (June 2005), pp. 5465-5471, ISSN: 0888-5885.

Gleick, P. H. (2001). *The World's water 2000-2001: The Biennial Report on Freshwater Resources*, Island Press, ISBN 1-55963-792-7, Washington, DC, USA.

Gleick, P. H. (2006). *The World's water 2006-2007: The Biennial Report on Freshwater Resources*, Island Press, ISBN 1-59726-108-8, Washington, DC, USA.

Gorenflo, A.; Velazques-Padron, D.; Frimmel, F. H. (2002) Nanofiltration of a German Groundwater of High hardness and NOM content: Performance and Costs, *Desalination*, Vol. 151, No. 3, (January 2003), pp. 253-265, ISSN: 0011-9164.

Greenlee, L. F.; Lawler, D. F.; Freeman, B. D.; Marrot, B.; Moulin, P. (2009) Reverse Osmosis Desalination: Water sources, Technology, and Today's Challenges, *Water Research*, Vol. 43, No. 9, (May 2009), pp. 2317-2348, ISSN: 0043- 1354.

Hadi, Khaled M. B. (2002) Evaluation of the Suitability of Groundwater Quality for Reverse Osmosis Desalination, *Desalination*, Vol. 142, No. 3, (March 2002), pp. 209-219, ISSN: 0011-9164.

Hamed, O. A.; Hassan, A. M.; Al-Shail, K.; Farooque, M. A. (2009) Performance Analysis of a Trihybrid NF/RO/MSF Desalination Plant, *Desalination and Water Treatment*, Vol. 1, No. 1-3, (January 2009), pp. 215-222, ISSN: 1944-3994.

Hamrouni, B.; Dhahbi, M. (2001) Thermodynamics Description of Saline Waters-Prediction of Scaling Limits in Desalination Processes, *Desalination*, Vol. 137, No. 1-3, (May 2001), pp. 275-284, ISSN: 0011-9164.

Hamrouni, B.; Dhabi, M. (2001) Analytical Aspects of Silica in Saline Waters- Application to Desalination of Brackish Waters, *Desalination*, Vol. 136, No. 1-3, (May 2001), pp. 225-232, ISSN: 0011-9164.

Hassan A. M., et al. (1998) A New Approach to Membrane and Thermal Seawater Desalination Processes using Nanofiltration membranes (Part1), *Desalination*, Vol. 118, No. 1-3, (September 1998), pp. 35-51, ISSN: 0011-9164.

Harkare, W. P., et al. (1982) Desalination of Brackish Water by Electrodialysis, *Desalination*, Vol. 42, No. 1, (July 1982), pp. 97-105, ISSN: 0011-9164.

International Desalination Association, (2009). Total World Desalination Capacity close to 60 million m³/d. In The *international Desalination and water Reuse Quarterly Industry Newsletter*, accessed 15.08.11, available from:
http://www.desalination.biz/news/news_story.asp?id=5121

Karagiannis, I. C.; Soldatos, P. G. (2008) Water Desalination Cost Literature: Review and Assessment, *Desalination*, Vol. 223, No. 1-3, (March 2008), pp. 448-456, ISSN: 0011-9164.

Katz, W. E. (1979) The Electrodialysis Reversal (EDR) Process, *Desalination*, Vol. 28, No. 1, (January 1979), pp. 31-40, ISSN: 0011-9164.

Katz, W. E. (1982) Desalination by ED and EDR – State-of-the-Art in 1981, *Desalination*, Vol. 42, No. 2, (August 1982), pp. 129-139, ISSN: 0011-9164.

Kedem, O.; Zalmon, G. (1997) Compact Accelerated Precipitation Softening (CAPS) as a Pretreatment for Membrane Desalination I. Softening by NaOH, *Desalination*, Vol. 113, No. 1, (November 1997), pp. 65-71, ISSN: 0011-9164.

Khawaji, A. D.; Kutubkhanah, I. K.; Wie, J. (2008) Advances in Seawater Desalination Technologies, *Desalination*, Vol. 221, No. 1-3, (March 2008), pp. 47-69, ISSN: 0011-9164.

Kim, D. H. (2011) A Review of Desalting Process Techniques and Economic Analysis of the Recovery of Salts from Retenates, *Desalination*, Vol. 270, No. 1-3, (April 2011), pp. 1-8, ISSN: 0011-9164.

Lee, K. P.; Arnot, T. C.; Mattia, D. (2011) A Review of Reverse Osmosis Mambrane Material for Desalination – Development to Date and Future Potential, *Journal of Membrane Science*, Vol. 370, No. 1-2, (March 2011), pp. 1-22, ISSN: 0376-7388.

Lee, S.; et al. (2009) Toward a Combined System of Forward Osmosis and Reverse Osmosis for Seawater Desalination, *Desalination*, Vol. 247, No. 1-3, (October 2009), pp. 239-246, ISSN: 0011-9164.

Lindsay, F. K.; Ryznar, J. W. (1939) Removal of Silica from Water by Sodium Aluminate, *Industrial and Engineering Chemistry*, Vol. 31, No. 7, (July 1939), pp. 859-861, ISSN: 0888-5885.

Mabrouk, A. A.; Nafey, A. S.; Fath, H. E. S. (2007) Analysis of a New Design of a Multi-Stage-Flash-Mechanical Vapor Compression Desalination Process, *Desalination*, Vol. 204, No. 1-3, (February 2007), pp. 482-500, ISSN: 0011-9164.

Mace, R. E.; Nicot, J.; Chowdhury, A. H.; Dutton, A. R.; Kalaswad, S., (2006), Please Pass the Salt: Using Oil Fields for the Disposal of Concentrate from Desalination Plants, Texas Water Development Board, Report No. 366, April 2006.

Masarawa, A.; Meyerstein, D.; Daltrophe, N.; Kedem, O. (1997) Compact Accelerated Precipitation Softening (CAPS) as a Pretreatment for Membrane Desalination II. Lime Softening with Contaminant Removal of Silica and Heavy Metals, *Desalination*, Vol. 113, No. 1, (November 1979), pp. 73-84, ISSN: 0011-9164.

Mathioulakis, E.; Belessiotis, V.; Delyannis, E. (2007) Desalination by Using Alternative Energy: Review and State-of-the-Art, *Desalination*, Vol. 203, No. 1-3, (February 2007), pp. 346-365, ISSN: 0011-9164.

Martinetti, C. R.; Childress, A. E.; Cath, T. Y. (2009) High Recovery of Concentrated RO brines using Forward Osmosis and Membrane Distillation, *Journal of Membrane Science*, Vol. 331, No. 1-2, (April 2009), pp. 31-39, ISSN: 0376-7388.

McCutcheon J. R.; McGinnis, R. L.; Elimelech, M. (2005a) A novel Ammonia-Carbon Dioxide Forward (Direct) Osmosis Desalination Process, *Desalination*, Vol. 174, No. 1, (April 2005), pp. 1-11, ISSN: 0011-9164.

McCutcheon J. R.; McGinnis, R. L.; Elimelech, M. (2005b) Desalination by Ammonia-Carbon Dioxide Forward Osmosis: Influence of Draw and Feed Solution Concentrations on Process Performance, *Journal of Membrane Science*, Vol. 278, No. 1-2, (July 2006), pp. 114-123, ISSN: 0376-7388.

McGinnis, R. L.; Elimelech, M. (2007) Energy Requirements of Ammonia-Carbon Dioxide Forward Osmosis Desalination, *Desalination*, Vol. 207, No. 1-3, (March 2007), pp. 370-382, ISSN: 0011-9164.

Mickley, M. C. (2001) Membrane Concentrate Disposal: Practice and Regulations. U.S. Department of the Interior, Bureau of Reclamation, Mickley & Associates.

Mickley, M. C., (2006) Membrane Concentrate Disposal: Practice and Regulation, Desalination and Water Purification Research and Development Program Report No. 123 2nd Ed., April 2006 (Agreement No. 98-FC-81-0054), Prepared for the U.S. Department of the Interior, Bureau of Reclamation, Technical Service Center, Water Treatment Engineering and Research Group.

Missimer T.M. (1994) Groundwater as a Feedwater Source for Membrane Treatment Plants: Hydrogeologic Controls on Water Quality Variations with Time, *Desalination*, Vol. 98, No. 1-3, (September 1994), pp. 451-457, ISSN: 0011-9164.

Mohamed, A. M. O.; Maraqa, M.; Al-Handaly, J (2005) Impact of Land Disposal of Reject Brine from Desalination Plant on Soil and Groundwater, *Desalination*, Vol. 182, No. 1-3, (November 2005), pp. 411-433, ISSN: 0011-9164.

Mohammadesmaeili, F.;Badr, M. K.; Abbaszadegan, M.; Fox, P. (2010) Byproduct Recovery from Reclaimed Water Reverse Osmosis Concentrate Using Lime and Soda-Ash

Treatment, *Water Environment Research,* Vol. 82, No. 4, (April 2010), pp. 342-350, ISSN: 1061-4303.

Neofotistou, E.; Demadis, K. D. (2004) Use of antiscalents for Mitigation of Silica (SiO_2) Fouling and Deposition: Fundamentals and Applications in Desalination Systems, *Desalination,* Vol. 167, (August 2004), pp. 257-272, ISSN: 0011-9164.

Nicot, J.; Chowdhury, A. H. (2005) Disposal of Brackish water Concentrate into Depleted Oil and Gas Fields: a Texas Study, *Desalination,* Vol. 181, No. 1-3, (September 2005), pp. 61-74, ISSN: 0011-9164.

Ning, R. Y. (2002) Discussion of Silica Speciation, Fouling, Control and Maximum Reduction, *Desalination,* Vol. 151, No. 1, (January 2003), pp. 67-73, ISSN: 0011-9164.

Ning, R. Y.; Tarquin, A.; Trzcinski, M. C.; Patwardhan, G. (2006) Recovery Optimization of RO Concentrate from Desert Wells, *Desalination,* Vol. 201, No. 1-3, (November 2006), pp. 315-322, ISSN: 0011-9164.

Ning, R. Y.; Troyer, T. L. (2009) Tandom Reverse Osmosis Process for Zero-Liquid Discharge, *Desalination,* Vol. 237, No. 1-3, (February 2009), pp. 238-242, ISSN: 0011-9164.

Oren, Y.; Katz, V.; Daltrophe, N. (2001) Improved Compact Accelerated Precipitation Softening (CAPS), *Desalination,* Vol. 139, No. 1-3, (September 2001), pp. 155-159, ISSN: 0011-9164.

Oren, Y; et al, (2010) Pilot Studies on High Recovery BWRO-EDR for Near Zero Liquid Discharge Approach, *Desalination,* Vol. 261, No. 3, (October 2010), pp. 321-330, ISSN: 0011-9164.

Phillip, W. A.; Yong, J. S.; Elimelech, M. (2010) Reverse Draw Solute Permeation in Forward Osmosis: Modeling and Experiments, *Environmental Science and Technology,* Vol. 44, No. 13, (July 2010), pp. 5170-5176, ISSN: 0013-936X.

Rahardianto, A.; Gao, J.; Gabelich, C. J.; Williams, M. D.; Cohen, Y. (2007) High Recovery Membrane Desalting of Low-Salinity Brackish Water: Integration of Accelerated Precipitation Softening with Membrane RO, *Journal of Membrane Science,* Vol. 289, No. 1-2, (February 2007), pp. 123-137, ISSN: 0376-7388.

Rahardianato, A; Mccool, B. C.; Cohen, Y. (2008) Reverse Osmosis Desalting of Inland Brackish water of High Gypsum Scaling Propensity: Kinetics and Mitigation of Membrane Mineral Scaling, *Environmental Science and Technology,* Vol. 42, No. 12, (June 2008), pp. 4292-4297, ISSN: 0013-936X.

Rautenbach, R; Groschl, A. (1990) Separation Potential of Nanofiltration membranes, *Desalination,* Vol. 77, (March 1990), pp. 73-84, ISSN: 0011-9164.

Reverter, J. A; Talo, S.; Alday, J. (2001) Las Palmas III – the Success Story of Brine Staging, *Desalination,* Vol. 138, No. 1-3, (September 2001), pp. 207-217, ISSN: 0011-9164.

Reynolds T. D.; Richards P. A., (1996), *Unit Operations and Processes in Environmental Engineering,* PWS Publishing Company, ISBN 0-534-94884-7, Boston, Massachusetts, USA.

Sadrzadeh, M.; Mohammadi, T. (2008) Sea Water Desalination Using Electrodialysis, *Desalination,* Vol. 221, No. 1-3, (March 2008), pp. 440-447, ISSN: 0011-9164.

Saitua, H.; Gil, R.; Padilla, A. P. (2011) Experimental Investigations on Arsenic Removal with a Nanofiltration Pilot Plant from Contaminated Groundwater, *Desalination*, Vol. 274, No. 1-3, (July 2011), pp. 1-6, ISSN: 0011-9164.

Schaep, J; et al. (1998) Removal of Hardness from Groundwater by Nanofiltration, *Desalination*, Vol. 119, No. 1-3, (September 1998), pp. 295-301, ISSN: 0011-9164.

Semiat, R.; Sutzkover, I.; Hasson, D. (2003) Scaling of RO Membranes from Silica Supersaturated Solutions, *Desalination*, Vol. 157, No. 1-3, (August 2003), pp. 169-191, ISSN: 0011-9164.

Semiat, R.; Hasson, D.; Zelmanov, G.; Hemo, I. (2004) Threshold Scaling Limits of RO Concentrates Flowing in a Long Waste Disposal Pipeline, *Water Science and Technology*, Vol. 49, No. 2, pp. 211-219, ISSN: 0273-1223.

Seto, T., et al. (1978) Seawater Desalination by Electrodialysis, *Desalination*, Vol. 25, No. 1, (March 1978), pp. 1-7, ISSN: 0011-9164.

Sheikholesami, R.; Tan, S. (1999) Effects of Water Quality on Silica Fouling of Desalination plants, *Desalination*, Vol. 126, No. 1-3, (November 1999), pp. 267-280, ISSN: 0011-9164.

Sheikholesami, R.; Al-Mutaz, I. S.; Koo, T.; Young, A. (2001) Pretreatment and the Effect of Cations and Anions on Prevention of Silica Fouling, *Desalination*, Vol. 139, No. 1-3, (September 2001), pp. 83-95, ISSN: 0011-9164.

Sheikholesami, R; Bright, J. (2002) Silica and Metals Removal by Pretreatment to Prevent Fouling of Reverse Osmosis Membranes, *Desalination*, Vol. 143, No. 3, (June 2002), pp. 255-267, ISSN: 0011-9164.

Sheikholesami, R. (2003a) Mixed salts – Scaling limits and Propensity, *Desalination*, Vol. 154, No. 2, (April 2003), pp. 117-127, ISSN: 0011-9164.

Sheikholesami, R. (2003b) Nucleation and Kinetics of Mixed Salts in Scaling, *AIChE Journal*, Vol. 49, No. 1, (January 2003), pp. 194-202, ISSN: 001-1541.

Sheikholesami, R. (2004) Assessment of the Scaling Potential for Sparingly Soluble Salts in RO and NF units, *Desalination*, Vol. 167, (August 2004), pp. 247-256, ISSN: 0011-9164.

Sheikholesami, R. (2009) Strategies for Future Research and Development in Desalination-challenges ahead, *Desalination*, Vol. 248, No. 1-3, (November 2009), pp. 218-224, ISSN: 0011-9164.

Sherif, M.; Mohamed, M.; Kacimov, A.; Shetty, A. (2011) Assessment of Groundwater Quality in the Northeastern Coastal area of UAE as Precursor for Desalination, *Desalination*, Vol. 273, No. 2-3, (June 2011), pp. 436-446, ISSN: 0011-9164.

Shon, H. K.; et al. (2009) Physicochemical Pretreatment of Seawater: Fouling Reduction and Membrane Characterization, *Desalination*, Vol. 238, No. 1-3, (November 2009), pp. 10-21, ISSN: 0011-9164.

Skehan, S; Kwiatkowski P. J. (2000) Concentrate Disposal via Injection Wells – Permitting and Design Considerations, *Florida Water Resources Journal*, Vol. 52, No. 5, (May 2000), pp. 19-21, ISSN: 0896-1794.

Tahaikt, M., et al. (2007) Fluoride Removal from Groundwater by Nanofiltration, *Desalination*, Vol. 212, No. 1-3, (June 2007), pp. 46-53, ISSN: 0011-9164.

Tait, S.; Clarke, W. P.; Keller, J.; Batstone, D. J. (2009) Removal of Sulfate from High-Strength Wastewater by Crystallization, *Water Research*, Vol. 43, No. 3, (February 2009), pp. 762-772, ISSN: 0043-1354.

Tang, W.; Ng, H. Y. (2008) Concentration of Brine by Forward Osmosis: Performance and influence of membrane structure, *Desalination*, Vol. 224, No. 1-3, (April 2008), pp. 143-153, ISSN: 0011-9164.

The Federal Institute for Geosciences and Natural Resources, (n.d.). Groundwater Resources of the World, accessed 15.08.11, available from:
http://www.bgr.bund.de/EN/Themen/Wasser/Bilder/Was_wasser_startseite_g w_erde_g_en.html?nn=1548136

United Nations UN, (1997). UN Assessment of Freshwater Resources 1997, accessed 13.08.11, available from:
http://www.un.org/ecosocdev/geninfo/sustdev/waterrep.htm

United States Environmental Protection Agency, US EPA, (2009) National Primary & Secondary Drinking Water Regulations, accessed 15.08.11, available at:
http://water.epa.gov/drink/contaminants/index.cfm

United Nations Environmental Programme UNEP- GRiD, (n.d.). Increased Global Water Stress, accessed 15.08.11, available from:
http://www.grida.no/graphic.aspx?f=series/vg-water2/0400-waterstress-EN.jpg

Valcour, H. C. (1985) Recent Applications of EDR, *Desalination*, Vol. 54, pp. 163-183, ISSN: 0011-9164.

Vandercaseele, C.; Van Der Bruggen, B. (2002) Distillation vs. Membrane Filtration: Overview of Process Evolutions in Seawater Desalination, *Desalination*, Vol. 143, No. 3, (June 2002), pp. 207-218, ISSN: 0011-9164.

Vince, F.; Marechal, F.; Aoustin, E.; Bréant, P. (2008) Multi-Objective Optimization of RO Desalination Plants, *Desalination*, Vol. 222, No. 1-3, (March 2008), pp. 96-118, ISSN: 0011-9164.

Walha, K.; Ben Amar, A.; Firdaous, L.; Quemeneur, F.; Jaouen, P. (2007) Brackish Groundwater Treatment by nanofiltration, Reverse Osmosis, and Electrodialysis in Tunisia: Performance and cost comparison, *Desalination*, Vol. 207, No. 1-3, (March 2007), pp. 95-106, ISSN: 0011-9164.

Wang, L. K (2008), *Handbook of Environmental Engineering V.13: Membrane and Desalination Technologies*, Humana Press, ISBN 1-58829-940-6, New York, USA.

WHO (2011) Guidelines for Drinking-Water Quality, 4th Ed. Accessed 15.08.11, available from: http://water.epa.gov/drink/contaminants/index.cfm

Yangali-Quintanilla, V.; Li, Z.; Valladares, R.; Li, Q.; Amy G. (2011) Indirect Desalination of Red Sea Water with Forward Osmosis and Low Pressure Reverse Osmosis for water reuse, *Desalination*, doi:10.1016/j.desal.2011.06.066.

Yun, T. I; Gabelich, C. J.; Cox, M. R.; Mofidi, A. A.; Lesan, R. (2006) Reducing Costs for Large-Scale Desalting Plants Using Large-Diameter Reverse Osmosis Membranes, *Desalination*, Vol. 189, No. 1-3, (March 2006), pp. 141-154, ISSN: 0011-9164.

Zubari, W. K. (2003) Assessing the Sustainability of Non-renewable Brackish Groundwater in Feeding an RO Desalination Plant in Bahrain, *Desalination*, Vol. 159, No. 3, (November 2003), pp. 211-244, ISSN: 0011-9164.

Zubari, W. K.; Madany, I. M.; Al-Junaid, S. S.; Al-Manaii, S. (1994) Trends in the quality of Groundwater in Bahrain with Respect to Salinity, 1941-1992., *Environment International*, Vol. 20, No. 6, pp. 739-746, ISSN: 0160-4120.

Part 2

Separation Technology

7

Organic/Inorganic Nanocomposite Membranes Development for Low Temperature Fuel Cell Applications

Touhami Mokrani
University of South Africa
South Africa

1. Introduction

The criteria that are going to influence the evolution of the world energy system in the present century are complex. The most important new factor is the need to preserve the environment, both locally and globally, through the use of new technologies and sustainable use of existing resources. The Kyoto protocol, which put a limit on greenhouse gas emissions (mainly CO_2) from the industrialized countries, is a turning point in the global energy chain. On the other hand, the fuel specifications to control automotive exhaust gas emission obligate fuel producers to look for different ways of making clean fuel. Automakers are also obligated to look for alternative technology to internal combustion engines. The interest in studies on energy sources alternative to fossil fuels is linked both to the reduction of their availability and the increasing environmental impact caused by their use (Goodstein, 1999). In the energy field, an important cause of pollutant emissions is linked to ground transportation. In the last 40 years, some economic, social and cultural changes have encouraged a wide proliferation of vehicles. For example, in Europe, private cars have increased from 232 to 435 per 1000 inhabitants in the period 1971-1995 (Santarelli *et al.*, 2003). Fuel cells are alternative power sources that can meet global emission regulations, and clean production. Although fuel cells have been used since the 1960's for aerospace and military applications, cost was a strong impediment to terrestrial applications.

2. Fuel cell types

Five major types of fuel cells are available and are defined by their electrolyte. These include alkaline (AFC), phosphoric acid (PAFC), molten carbonate (MCFC), solid oxide (SOFC) and proton exchange membrane fuel cells (PEMFC). Table 1 summarizes some characteristics of these fuel cells. Proton exchange membrane fuel cells are the most attractive candidate for alternative automotive and stationary power sources due to their smaller size and much lower operating temperature compared to other fuel cell systems. Low temperature fuel cells are fuel cells operating at temperature less than 100°C. They are H_2-proton exchange membrane fuel cell (H_2-PEMFC), direct methanol fuel cell (DMFC), direct ethanol fuel cell (DEFC) and direct DME fuel cell (DDMEFC).

Type	Electrolyte	Charge carrier in the electrolyte	Temperature (°C)
Alkaline fuel cells (AFC)	aqueous KOH solution	OH-	<100
Proton exchange membrane fuel cells (PEMFC)	proton exchange membrane	H+	60-120
Phosphoric acid fuel cells (PAFC)	concentrated phosphoric acid	H+	160-220
Molten carbonate fuel cells (MCFC)	mixture of molten carbonates (Li_2CO_3/K_2CO_3)	CO_3^{2-}	600-650
Solid oxide fuel cells (SOFC)	ceramic solid $ZrO_2(Y_2O_3)$	O_2^-	800-1000

Table 1. Fuel cells systems (Carrette *et al.*, 2001)

3. Low temperature fuel cells

A fuel cell is an electrochemical system which converts chemical energy to electrical energy. A fuel cell differs from a battery in that fuels are continuously supplied and the products are continuously removed. There are two distinct fuels for low temperature fuel cells: hydrogen as used in a H_2-PEMFC, and methanol as used in a DMFC. These fuel cells consist of six major parts: end plates, current collectors, flow channel blocks, gaskets, gas diffusion layers, and a membrane electrode assembly (MEA). The fuel cell principle enables a separation between power and energy. The maximum power required determines the size of the fuel cell; the energy required determines the amount of fuel to be carried. The specific power (W kg-1) of the H_2-PEMFC is roughly twice that of the DMFC (Raadschelders & Jansen, 2001). Because no mobile electrolyte is employed, corrosion problems in low temperature fuel cells are reduced and cell construction is simplified with few moving parts (Bernardi & Verbrugge, 1991). Also, fuel cells operate very quietly, therefore, reducing noise pollution (Kordesch & Simader, 1995). Since the proton exchange membrane used for the electrolyte is a solid phase, it does not penetrate deeply into the electrode as does a liquid one; therefore the reaction area is limited to the contact surface between the electrode and membrane (Shin *et al.*, 2002). The advantage of using solid electrolyte is that no electrolyte leakage will occur (Uchida *et al.*, 1995; Yi & Nguyen, 1999). To meet the requirements of practical application a large number of single cells are assembled together to form a stack. The performance of a stack is different from that of a single cell. The stack has a much higher operating voltage, a greater power and better fuel-energy efficiency (Chu & Jiang, 1999).

4. Fuels for low temperature fuel cells

4.1 Pure hydrogen

H_2-proton exchange membrane fuel cells have existed since the 1960's; in fact they were used in the Gemini aerospace program of the National Aeronautics and Space Administration (NASA) of the United States. The MEA for H_2-PEMFCs consists of five components namely: a porous backing layer, an anode catalyst layer, a proton exchange membrane, a cathode catalyst layer, and a porous backing layer. Hydrogen is oxidized at the anode. The proton formed migrates through the membrane while the electrons flow through the external circuit. In the cathode reaction water is formed from oxygen, protons and electrons.

The two half reactions for the H_2-PEMFC are as follows:

Oxidation half reaction: Anode $2 H_2 \longrightarrow 4H^+ + 4e^-$

Reduction half reaction: cathode $O_2 + 4H^+ + 4e^- \longrightarrow 2 H_2O$

_____

Cell reaction $2 H_2 + O_2 \longrightarrow 2 H_2O$

H_2-PEMFCs have attracted the most attention due to their high electrochemical reactivity (Gottesfeld & Zawodzinski, 1997; Parthasarathy et al., 1991; Ralph, 1997) and very low noble catalyst loading since the development of a method at Los Alamos National Laboratory (LANL) to reduce the platinum loading to ca. 0.1 mg/cm^2 (Wilson & Gottesfeld, 1992; Wilson, 1993; Wilson et al., 1995) compared to 35 mg/cm^2 and 4 mg/cm^2 used respectively in the Gemini program and at General Electric in the 1970s (Appleby & Yeager, 1986a,1986b). The efficiency achievable is higher than in power plants and internal combustion engines (Dohle et al., 2002) and there is practically zero pollution. However, the H_2-PEMFC has several disadvantages including hydrogen storage and transportation and the public acceptance of hydrogen as fuel. It is well known that hydrogen and air mixtures are explosive (e.g. the Challenger disaster). Hydrogen safety measures are still one of the major implications when it comes to the commercialization of H_2-PEMFCs. Adequate water content of the membranes is essential to maintain the conductivity of the polymeric proton exchange membrane (Anantaraman & Gardner, 1996; Fontanella et al., 1995; Gavach et al., 1989; Zawodzinski et al., 1991, 1993). During fuel cell operation, water molecules migrate through the membrane under electro-osmotic drag, fluid convection, and molecular diffusion, making it difficult to retain a high water content within the membrane. Generally, humidification is applied to the inlets of the anode and/or cathode in order to supply water to the membrane. However, excessive amounts of liquid water could impede mass transport within the electrode structure (Zawodzinski et al., 1991). A thinner membrane is preferred in H_2-PEMFCs because it can provide an improvement in water management due to the enhanced back-diffusion of production water from the cathode to the anode side (Finsterwalder & Hambitzer, 2001). The oxygen reduction reaction (ORR) is very slow compared to the hydrogen reaction; typically hydrogen electro-oxidation on Pt is shown by an exchange current density of 10^{-3} A cm^{-2} Pt at ambient temperature. This is some 10^7 to 10^9 times more facile than the oxygen reduction at the cathode (Ralph & Hogarth, 2002). Thus, oxygen reduction is a rate limiting factor in H_2-PEMFCs (Gloaguen et al., 1998; Paulus et al., 2001).

4.2 Hydrogen reformate

The question of whether customers will be fuelling their vehicles directly with hydrogen or via the hydrogen-rich carrier (e.g. methanol, ethanol, gasoline, diesel, etc.) still seems to be unanswered. This is a very important issue not just from a refueling infrastructure perspective but also from the public perception and from the gearing up of production, and developing guidelines for dealing with safety issues that will need to put in place for the new fuel (Adamson & Pearson, 2000). In principle, any type of liquid fuel may be employed as a hydrogen source, e.g. gasoline, diesel, methanol, ethanol, etc. Hydrogen is produced by a reforming process. Four distinguish fuels are discussed namely methanol, ethanol, dimethylether (DME) and ammonia.

4.2.1 Methanol reforming

Methanol is produced from steam reformed natural gas and carbon dioxide using copper-based catalyst, and also from renewable biomass sources. Methanol is a leading candidate to provide the hydrogen necessary to power a fuel cell, especially in vehicular applications (Ledjeff-Hey et al., 1998; Mokrani & Scurrell, 2009; Olah et al., 2009). Methanol is currently used as a feed stock for a variety of widely used organic chemicals, including formaldehyde, acetic acid, chloromethane, and methyl tert-butyl ether (MTBE). Methanol is the desired fuel to produce hydrogen on-board. Methanol can be reformed to hydrogen by different processes including steam reforming (Amphlett et al., 1985; Breen & Ross, 1999; Duesterwald et al., 1997; Emonts et al., 1998; C.J. Jiang et al., 1993; Takahashi et al., 1982; Takezawa et al., 1982), partial oxidation (Agrell et al., 2001; Cubiero & Fierro, 1998; Velu et al., 1999) and autothermal reforming (Edwards et al., 1998; Höhlein et al., 1996; L. Ma et al., 1996; Mizsey et al., 2001).

Steam reforming of methanol occurs by two different pathways (Emonts et al., 1998). The first one involves the decomposition of methanol into CO and H_2 through the following reaction:

$$CH_3OH \rightleftharpoons CO + 2H_2$$

followed by a water gas shift reaction:

$$CO + H_2O \rightleftharpoons CO_2 + H_2$$

The second mechanism for methanol steam reforming consists of the reaction of water and methanol to CO_2 and hydrogen:

$$CH_3OH + H_2O \rightleftharpoons CO_2 + 3H_2$$

which can be followed by a reverse shift reaction to establish the thermodynamic equilibrium:

$$CO_2 + H_2 \rightleftharpoons CO + H_2O$$

Methanol steam reforming is endothermic and therefore requires that external heat, typically 300°C, is supplied. Steam reforming of methanol is usually catalyzed over Cu/ZnO type catalyst and can be performed in fixed-bed reactors (Duesterwald et al., 1997).

4.2.2 Ethanol reforming

Among other candidate liquid fuels, ethanol is a particular case, since it can be easily produced in great quantity by the fermentation of sugar-containing raw materials. In addition, in some countries (e.g. Brazil) ethanol is already distributed in gas stations for use in conventional cars with internal combustion engines. Hydrogen is produced from ethanol in a process unit consisting of either a steam reformer (SR) or a partial oxidation (POX) reactor in series with a water-gas shift (WGS) reactor and a reactor for selective oxidation (PROX) of CO (Ioannides & Neophytides, 2000). Product gas from the reformer or the POX reactor, which operates at an exit temperature higher than 677°C, contains a mixture of H_2, CO, CO_2, CH_4 and H_2O. After cooling, this stream enters the WGS reactor, where a large fraction of CO reacts with H_2O towards CO_2 and H_2 at a temperature of 200°C. The product gas of the WGS reactor contains 0.1-1.5% of residual CO and enters the PROX reactor, where CO is totally oxidized - with the addition of a small amount of air - to CO_2 with residual CO being less than 10 ppm. The CO free, hydrogen rich stream is then fed to the H_2-PEMFC.

4.2.3 DME reforming

DME (dimethylether) has become a promising candidate as a hydrogen source for the reforming process, because it has a high hydrogen-to-carbon ratio and a high energy density. DME can be easily handled, stored and transported. Furthermore, the infrastructure of LPG can readily be adapted for DME due to their similar physical properties. Furthermore, DME is not toxic and less explosive. DME can be catalytically reformed at relatively lower temperatures than ethanol and methane. DME can be reformed through three ways, namely steam reforming (SR), partial oxidation (POX) and authothermal reforming (ATR). DME SR proceeds via two moderately endothermic reactions in sequence; hydrolysis of DME to MeOH and steam reforming of MeOH to hydrogen and carbon dioxide. Hydrolysis of DME takes place over acid catalysis, e.g. zeolite and alumina, while MeOH SR proceeds over Cu-, Pt-, or Pd based catalyst. Therefore, bi-functional catalyst containing both acidic and metallic sites are generally needed for DME SR (Faungnawakij et al., 2010; Ledesma & Llorca, 2009; Nishiguchi et al., 2006; Takeishi & Suzuki, 2004). DME POX has been investigated over various metal catalysts such as Pt, Ni, Co and Rh supported on different oxide. Supports such as Al_2O_3, YSZ, $LaGaO_3$-based and MgO were used at a high reaction temperature ranging from 400 to 700°C (S. Wang et al., 2002; Q. Zhang et al., 2005). ATR also can be used to produce hydrogen from DME. ATR is a combination of SR and POX, and catalysis such as $CuFe_2O_4$-Al_2O_3 (Faungnawakij & Viriya-empikul, 2010) and Pd-based (Nilsson et al., 2007, 2009) were investigated.

4.2.4 Ammonia reforming

Anhydrous ammonia is a widely used commodity and is available worldwide in liquid form in low pressure tanks. Procedures for safe handling have been developed in every country. Facilities for storage and transport by barges, trucks and pipelines from producer to ultimate consumer are available throughout the world. Therefore, liquid anhydrous ammonia is an excellent storage medium for hydrogen (Hacker & Kordesch, 2003). Studies demonstrate that hydrogen derived from anhydrous liquid ammonia, via a dissociation and followed by hydrogen purifier, offers an alternative to conventional methods of obtaining pure hydrogen for small scale use (Strickland, 1984). The dissociation rate depends on temperature, pressure and the catalyst being used. An almost complete decomposition of ammonia can take place at approximately 430°C at atmospheric pressure. The influence and kinetic data of materials like porcelain or silica glass, metals like ion, tungsten, molybdenum, nickel, etc. especially noble metals and metal oxides, have been investigated for the dissociation of ammonia. The most used catalysts are nickel oxide and iron oxide (Papapolymerou & Bontozoglou, 1997) and the better cracking efficiencies were obtained with catalysts based on Zr, Mn, Fe and Al/alloys (Boffito, 1999; Rosenblatt & Cohn, 1952; Shikada et al., 1991).

4.3 Direct methanol fuel cell

DMFC technology is relatively new compared to the H_2-PEMFC. However, the direct oxidation of methanol in a DMFC has been investigated over many years and some prototypes were built in the 1960's and early 1970's by the Shell Research Center in England (Glazebrook, 1982; Schatter, 1983) and by Hitachi Research Laboratories in Japan (Tamura et al., 1984; Williams, 1966). These studies were abandoned in the mid-1980's due to the low performance (25 mW cm^{-2} at best) resulting from the use of a liquid acid electrolyte (Glazebrook, 1982; Kordesch & Simader, 1996; Lamy et al., 2001). An alkaline electrolyte was also used, but evolved CO_2 caused carbonation of the electrolyte resulting in decreased efficiency by reducing the electrolyte conductivity and de-polarizing the cathode (Verma, 2000). Currently all the research in DMFCs focuses on using solid proton exchange membranes as electrolyte, largely due to its proliferation in H_2-PEMFCs. The structure of the DMFC is similar to the H_2-PEMFC. At the anode methanol is directly oxidized to carbon dioxide, and the reaction at the cathode is similar to the H_2-PEMFC.

The two main half reactions for the DMFC can be summarized as follows:

Oxidation half reaction: Anode $\quad CH_3OH + H_2O \longrightarrow CO_2 + 6H^+ + 6e^-$

Reduction half reaction: cathode $\quad 3/2\, O_2 + 6H^+ + 6e^- \longrightarrow 3\, H_2O$

Cell reaction $\quad\quad\quad\quad\quad\quad CH_3OH + 3/2\, O_2 \longrightarrow CO_2 + 2\, H_2O$

The thermodynamic reversible potential for a DMFC is 1.21V at 25°C (Larminie & Dicks, 2000). This value is comparable to that for a H_2-PEMFC, which is 1.23V (Chu & Gilman, 1994; Léger, 2001; Qi & Kaufman, 2002; Scott et al., 1998). In practice, a DMFC has a much lower open circuit voltage (OCV) (Qi & Kaufman, 2002) and electrochemical losses at both electrodes lead to a significant reduction in overall performance from the theoretical thermodynamic maximum (Argyropoulos et al., 1999a). Since methanol is used directly at the anode, and as a consequence, a DMFC requires less auxiliary equipment and is therefore a more simplified system compared to a H_2-PEMFC. Methanol is a liquid made from natural gas or renewable biomass sources, which is relatively cheap. Methanol is also easy to store, transport, and distribute, where advantage can be taken of the existing gasoline infrastructure (Mokrani & Scurrell, 2009; Olah et al., 2009). The anodic reaction is exothermic for both the H_2-PEMFC and the DMFC; heat management is a problem in H_2-PEMFC stacks. In contrast, aqueous methanol acts as a coolant in DMFCs (Hogarth et al., 1997; Hogarth & Ralph, 2002; Lim & C.Y. Wang, 2003; Surampudi et al., 1994).

However, as the DMFC is still in its infancy, many problems need to be overcome to reach the commercialization stage. This includes the very sluggish methanol oxidation reaction, methanol crossover through the polymeric proton exchange membrane, CO_2 evolvement at the anode (Argyropoulos et al., 1999a,1999b; Nordlund et al., 2002; Scott et al., 1998), and cathode flooding (Amphlett et al., 2001; M. Mench et al., 2001; X. Ren & Gottesfeld, 2001; Z.H. Wang et al., 2001). The methanol crossover through the polymer electrolyte leads to a mixed potential at the cathode, which results from the ORR and the methanol oxidation occurring simultaneously. This effect causes a negative potential shift at the cathode and a significant decrease of performance in the DMFC. Methanol crossover also causes fuel loses; it had been found that over 40% of methanol can be wasted in a DMFC across Nafion® membranes (Narayanan et al., 1996). In a DMFC, cathode flooding, which typically occurs unless high cathode stoichiometries are used, can determine to a great extent overall cell performance (Amphlett et al., 2001; M. Mench et al., 2001; X. Ren & Gottesfeld, 2001; Z.H. Wang et al., 2001). Water management in the DMFC is especially critical because anode water activity is near unity due to contact with liquid methanol solution (M.W. Mench & C.Y. Wang, 2003). Thus, unlike a H_2-PEMFC, no back-diffusive flux of water from cathode to anode will occur, and as a result, vapourization into dry cathode flow is the only pathway for removal of excess cathode-side water accumulation from electro-osmotic drag, ORR, and diffusion (M.W. Mench & C.Y. Wang, 2003).

4.4 Direct ethanol fuel cell

Direct fuel utilization will be of interest. Besides methanol, other alcohols, particularly those coming from biomass resources, are being considered as alternative fuels. Ethanol as an attractive fuel for electrical vehicles was investigated in direct ethanol fuel cells (Fujiwara et al., 1999; Gong et al., 2001; Lamy et al., 2001; W.J. Zhou et al., 2004). However, multimetallic catalysts are necessary to orientate the oxidation reaction selectively in the direction of complete combustion to carbon dioxide (Lamy et al., 2001). The reaction mechanisms of anodic oxidation of ethanol are more difficult to elucidate than methanol oxidation, since the number of electrons exchanged greatly increases (12 electrons per ethanol molecule versus 6 electrons for methanol), thus many adsorbed intermediates and products are involved (Lamy et al., 2001). Direct ethanol fuel cell was the second most studied fuel cell after methanol. The proton conductor membranes used are mainly Nafion® membranes. However some investigators used high temperature membrane such as Nafion®/Silica by Aricò et al. (1998), and also by J. Wang et al. (1995) using a phosphoric acid doped polybenzimidazol (PBI) membrane.

4.5 Direct DME fuel cell

Serov and Kwak (2009) have summarized the recent progress in development of direct DME fuel cell (DDMEFC). DDMEFC could be a valuable direct liquid fuel cell candidate for commercialization. However, compared with PEMFC and DMFC, DDMEFC show poor performances under ambient conditions due to the poor electrooxidation reactivity of DME (Colbow et al., 2000; Kerangueven et al., 2006; Mench et al., 2004; Mizutani et al., 2006; Ueda et al., 2006). On the anode side of DDMEFC, the following oxidation reaction takes place:

$$CH_3OCH_3 + 3H_2O \longrightarrow 2CO_2 + 12\,H^+ + 12\,e^-$$

The number of electron transferred for complete oxidation is 12, this result in a reduced theoretical fuel requirement of DME, when compared to methanol with 6 electrons transferred, and hydrogen with 2 electrons (Mench *et al.*, 2004; Serov & Kwak, 2009; K. Xu *et al.*, 2010). Furthermore, DME has the advantage over methanol in that crossover is much less pronounced (Mench *et al.*, 2004; Serov & Kwak, 2009). DDMEFC performance increase with increasing temperature, since DME electrooxidation is favored at high temperature. Furthermore, increasing the temperature will enhance also oxygen reduction reaction (ORR) (K. Xu *et al.*, 2010). Compared with hydrogen as the fuel for PEMFC, more water is needed for DME electrooxidation reaction and H^+ migration from anode side to cathode side due to the elctro osmotic force of water (Ferrell *et al.*, 2010; K. Xu *et al.*, 2010; Yu *et al.*, 2005).

5. Organic proton conductor membranes

Proton exchange membranes or proton conductor membranes are the most important component of low temperature fuel cells. Since the development of a solid polymer electrolyte, all the research on fuel cells focuses on the use of these types of electrolyte.

5.1 Perfluorinated membranes

The first commercially available perfluorinated membrane material from DuPont was Nafion® 120 (1200 equivalent weight (EW), 250 µm thick) followed by Nafion® 117 (1100 EW, 175 µm thick). These high equivalent weight materials were found to have limited use in fuel cells. In 1988, The Dow Chemical Company developed their own perfluorinated polymer membrane with low equivalent weight, typically in the range of 800-850. Nafion® of DuPont and Dow® membranes have identical backbones and are structurally and morphologically similar, but the side chain is shorter in the Dow polymer. Since the success of Dow Chemical, where it was found that the Dow® membrane performed better than the DuPont membrane in H_2/O_2 fuel cells, DuPont has been active in further developing their membranes with respect to durability and continuous improvement. They increased power densities by further decreasing the equivalent weight from 1100 to 1000 EW and membrane thickness from 175 to 25 µm. Table 2 shows the latest DuPont membranes with some characteristics. Nafion® 117 is the preferred membrane for DMFCs. In the 1990's, Aciplex® perfluorinated ion exchange membranes were introduced by the Asahi Chemical Industry, and the Flemion® series were introduced by Asahi Glass Co. (Yoshida *et al.*, 1998). In general these membranes are in the category of long chain perfluorinated membranes, like Nafion®. Some characteristics of these perfluorinated membranes are summarized in Table 2.

Nafion® membranes are chemically synthesized in four steps according to the DuPont de Nemours process (Grot, 1978): 1) The reaction of tetrafluoroethylene with SO_3 to form the sulfone cycle; 2) The condensation of these products with sodium carbonate followed by co-polymerization with tetrafluoroethylene to form an insoluble resin; 3) The hydrolysis of this resin to form a perfluorosulfonic polymer and 4) The chemical exchange of the counter ion Na^+ with the proton in an appropriate electrolyte. The Dow® membrane is prepared by the co-polymerisation of tetrafluoroethylene with vinylether monomer. The polymer can be described as having a Teflon-like backbone structure with a side chain attached via an ether group. This side chain is characterized by a terminal sulfonate functional group (Savadogo, 1998).

Membrane	Thickness (µm)	Equivalent Weight
Nafion® series (DuPont)		
Nafion® 117	175	1100
Nafion® 115	125	1100
Nafion® 112	50	1100
Nafion® 111	25	1100
Nafion® 1135	87	1100
Nafion® 1035	87	960
Nafion® 105	125	960
Dow Chemicals Co.		
Dow® XUS 13204.10	127	800–850
Flemion® series (Asahi Glass Co.)		
Flemion® R	50	900
Flemion® S	80	900
Flemion® T	120	900
Aciplex® series (Asahi Chemicals Industry)		
Aciplex® 1004	100	1000

Table 2. Perfluorinated membranes

5.2 Partially fluorinated ionomer membranes

5.2.1 Sulfonated copolymer based on the α,β,β-trifluorostyrene monomer membranes

The Canadian Ballard company developed proton conductor membranes based on trifluorostyrene monomer, under the trade name BAM1G and BAM2G (Ballard Advanced Materials first and second generation, respectively). The longevity of these polymers was limited to approximately 500 hours under practical fuel cell operating conditions (Savadogo, 1998). Based on the above work, Ballard developed third generation membranes under the trade name BAM3G (Steck, 1995; Steck & Stone, 1997; Wei et al., 1995a,1995b). The BAM3G membranes consist of sulfonated copolymers incorporating α,β,β-trifluorostyrene and a series of substituted α,β,β-trifluorostyrene co-monomers. These membranes have an equivalent weight ranging between 375 and 920. The water content of the sulfonated BAM3G is much higher than that of Nafion® and Dow membranes. BAM3G membranes demonstrated a lifetime approaching 15,000 hours when tested in a Ballard MK5 single cell and also exhibited performances superior to Nafion® and Dow® membranes in a H_2/O_2 fuel cell. Disadvantages of these membranes include the complicated production process for the monomer α,β,β-trifluorostyrene (Livingston et al., 1956) and the difficult sulfonation procedure (Kerres, 2001; Wei et al., 1995b).

5.2.2 Grafted ionomer membranes

Partially fluorinated membranes can be obtained by using a simultaneous and pre-radiation grafting of monomers onto a base polymer film, and subsequent sulfonation of the grafted component (Brack et al., 2003; Büchi et al., 1995a,1995b; Gode et al., 2003; Gupta et al., 1993; Hatanaka et al., 2002; W. Lee et al., 1996; Lehtinen et al., 1998; Scherer, 1990). These membranes were prepared by pre-irradiation of fluoropolymer films, such as poly(tetrafluoroethylene-co-hexafluoropropylene) (FEP) or poly(ethylene-alt-tetrafluoro-ethylene) (ETFE), using an electron beam or gamma irradiation source. The pre-irradiated films were grafted by exposing them to solutions of styrene and other radically polymerizable monomers. The grafted films are sulfonated using chlorosulfonic acid. The grafting mixture was crosslinked with divinylbenzene (DVB) and tri-allyl cyamirate (TAC) (Gupta et al., 1994; Gupta & Scherer, 1994) or poly(vinylidene fluoride) (Sundholm, 1998). A disadvantage of membranes using styrene and divinylbenzene monomers is that their oxidation stability is limited, due to the tertiary C-H bonds which are sensitive to O_2 and hydrogen peroxide attack (Kerres, 2001).

5.3 Non-perfluorinated membranes

5.3.1 Polybenzimidazole (PBI)

PBI is synthesized from aromatic bis-o-diamines and dicarboxylates (acids, esters, amides), either in the molten state or in solution (Jones & Rozière, 2001). PBI is relatively low cost and is a commercially available polymer known to have excellent oxidation and thermal stability. The commercially available polybenzimidazol is poly-[2,2`-(m-phenylene)-5,5`-bibenzimidazole], which is synthesized from diphenyl-iso-phthalate and tetra-aminobiphenyl. Hoel and Grunwald (1977) reported on proton conductivity values of PBI in the range of $2 \times 10^{-4} - 8 \times 10^{-4}$ S/cm at relative humidities (RH) between 0 and 100%. Other authors observed proton conductivity some two to three orders of magnitude lower (Aharoni & Litt, 1974; Glipa et al., 1997; Powers & Serad, 1986). PBI is a suitable basic polymer which can readily be complexed with strong acids (Jones & Rozière, 2001; Glipa et al., 1997; Y.L. Ma et al., 2004; Powers & Serad, 1986; Samms et al., 1996; Savadogo & B. Xing, 2000; Wainright et al., 1995, 1997; J.T. Wang et al., 1996a,1996b; B. Xing & Savadogo, 1999). The immersion of PBI film in aqueous phosphoric acid leads to an increase in both its conductivity and thermal stability (Powers & Serad, 1986). Savinell and co-workers prepared PBI/H_3PO_4 via two different routes: a) directly casting a film of PBI from a solution containing phosphoric acid; b) preparation by immersion of a preformed PBI membrane in 11M phosphoric acid for several days (Samms et al., 1996; Wainright et al., 1997). The typical thickness for different films was 75 μm. The conductivity depends on the quantity of phosphoric acid in the membrane. Conductivity in the range 5×10^{-3} to 2×10^{-2} S/cm at 130°C and 5×10^{-2} S/cm at 190°C have been reported (Wainright et al., 1995). The conductivity for type "a" membranes is higher than those of type "b" membranes. At a temperature above 150°C, the conductivity of type "a" membranes is similar to that of Nafion® at 80 °C and 100% RH. It was shown that the methanol crossover through doped PBI type "a" membrane, was at least ten times less than that observed with Nafion®. The disadvantage of these membranes is that the H_3PO_4 molecules can diffuse out of the membrane towards basic polymer sites because they are in excess. PBI/H_3PO_4 membranes are suitable for direct methanol fuel cell application at a temperature >100°C. However, they can only be used with a feed of vapourized methanol, because when a liquid contacts the membrane, the phosphoric acid leaches out of the membrane and the proton conductivity drops considerably (Kerres, 2001).

5.3.2 Sulfonated polyimide membranes

The sulfonated polyimide (SPI) membranes were obtained by casting on a glass plate the polymer solution and evaporating the solvent (Cornet *et al.*, 2000; Fan *et al.*, 2002; Faure *et al.*, 1996, 1997; Gebel *et al.*, 1993; Genies *et al.*, 2001; Woo *et al.*, 2003). The polymer solution synthesis was achieved in different ways: The first way was based on the phthalimide-five member imide (4,4'-diamino-biphenyl 2,2' disulfonic acid (BDSA), 4,4' oxy-diphthalic dianhydride (ODPA) and 4,4'–oxydianiline (ODA)) at 200°C. The second way was based on the naphthalimide-six member imide ring (BDSA, 1,4, 5,8-naphthalene tetracarboxylic dianhydride (NTDA) and ODA) at 160°C (Faure *et al.*, 1996, 1997; Gebel *et al.*, 1993). The third way was based on the 3,3`,4,4`-benzophenone-tetracarboxylic dianhydride (BTDA), BDSA and ODA (D`Alelio, 1944). The fourth way was based on BDSA/NTDA/mAPI (bis-[3-(Aminophenoxy)-4-phenyl]isopropylidene) (Genies *et al.*, 2001). The water content of membranes at 25°C for the phthalic and naphthalenic sulfonated polyimide membranes is 26% and 30%, respectively. The water content obtained for Nafion® membranes under the same conditions was 20% (Faure *et al.*, 1997; Gebel *et al.*, 1993). It was also claimed that the sulfonated polyimide membranes were 3 times less permeable to hydrogen gas than Nafion® membranes. The lifetime measurements were performed on a 175 μm phthalic polyimide and a 70 μm naphthalenic sulfonated polyimide film at 60°C, 3 bar pressure for H_2 and O_2 and under a constant current density. It was found that the membrane based on the phthalic structure broke after 70 hours whereas the membrane based on the naphthalic polyimide was stable over 3000 hours (D`Alelio, 1944). The proton conductivity of SPI was found to be half of Nafion® 117, typically 4.1×10^{-2} S/cm, and methanol permeability was found to be 7.34×10^{-8} compared to 2.38×10^{-6} cm²/s for Nafion® 117 (Woo *et al.*, 2003).

5.3.3 Phosphazene-based cation-exchange membranes

It was shown that polyphosphazene-based cation-exchange membranes have a low methanol permeability, low water swelling ratios, satisfactory mechanical properties, and a conductivity comparable to that of Nafion® 117 (Allcock *et al.*, 2002a,2002b; Guo *et al.*, 1999; Tang *et al.*, 1999; Wycisk & Pintauro, 1996; X. Zhou *et al.*, 2003). Polyphosphazene-based membranes have been fabricated from poly[bis(3-methylphenoxy)phosphazene] by first sulfonating the base polymer with SO_3 and then solution-casting a thin film (Tang *et al.*, 1999; Wycisk & Pintauro, 1996; X. Zhou *et al.*, 2003). Polymer crosslinking was carried out by dissolving benzophenone photoinitiator in the membrane casting solution and then exposing the resulting films after solvent evaporation to UV light (Guo *et al.*, 1999). The conductivity of the polyphosphazene membranes were either similar to or lower than that of Nafion® 117 membranes (Guo *et al.*, 1999; X. Zhou *et al.*, 2003). However, methanol permeability of a sulfonated membrane was about 8 times lower than that of the Nafion® 117 membrane (X. Zhou *et al.*, 2003). Sulfonated/crosslinked polyphosphazene films showed no signs of mechanical failure (softening) up to 173°C and a pressure of 800 kPa (Guo *et al.*, 1999).

5.3.4 Sulfonated poly(arylethersulfone) membranes

Polysulfone (PSU) is a low cost, commercially available polymer (e.g. PSU Udel™ from Amoco) which has very good chemical stability. The synthesis and characterization of sulfonated polysulfone (SPSU) has been achieved by Johnson et al. (1984) and Nolte et al. (1993). It was found that membranes cast from SPSU (Udel™ P-1700) solutions were completely water soluble (Nolte et al., 1993) and become very brittle when drying out which can happen in the fuel cell application under intermittent conditions (Kerres et al., 1999).

There are two new but different procedures for the sulfonation of polysulfone. In one procedure, the sodium-sulfonated group was introduced in the base polysulfone via the metalation-sulfination-oxidation process (Kerres et al., 1996, 1998a). In the other procedure, trimethylsilyl chlorosulfonate was used as the sulfonating agent (Baradie et al., 1998). Lufrano et al. (2000,2001) prepared SPSU via trimethylsilyl chlorosulfonate with different degrees of sulfonation (DS). Different membranes with sulfonation degree from 23% to 53% (Lufrano et al., 2000) on the one hand and 49%, 61% and 77% (Lufrano et al., 2001) on the other hand were prepared. With a 61% sulfonation degree a proton conductivity for SPSU of 2.7×10^{-2} S/cm at 25°C was reported (Lufrano et al., 2001). This conductivity was 3.5 times lower than Nafion® 117, but was compensated by the lower thickness, 90 μm vs. 210 μm for Nafion® 117. The cell performance obtained by Lufrano et al. (2001) was almost the same for SPSU and Nafion® in a H_2/O_2 fuel cell. This is higher than that reported previously by Kerres et al. (1998a) and Baradie et al. (1998). Y.S. Kim et al. (2003) prepared sulfonated poly(arylether sulfone) membranes.

Promising alternatives suggested by Kerres and co-workers, include composite membranes made from blends of acidic and basic polymers (Cui et al., 1998; Jörissen et al., 2002; Kerres et al., 1999, 2000; Walker et al., 1999) or modified PSU via the metalation-sulfochlorination and the metalation-amination routes (W. Zhang et al., 2001) or crosslinked SPSU (Kerres et al., 1997, 1998b,1998c). These alternatives are made by blending acidic polymers such as SPSU with basic polymers such as poly(4-vinylpyridine) (P4VP), polybenzimidazole (PBI) or a basically substituted polysulfone. Crosslinked SPSU blend membranes have been produced via a new crosslinking process. The blends have been obtained by mixing PSU Udel™ Na-sulfonate and PSU Udel™ Li-sulfinate in N-methyl pyrrolidone. The membranes have been crosslinked by S-alkylation of PSU sulfinate groups with di-halogenoalkanes. These membranes show very good performance in H_2/O_2 fuel cells and DMFCs (Kerres et al., 2000; Kerres, 2001). These membranes also show a markedly reduced methanol permeability (Kerres, 2001; Walker et al., 1999).

5.3.5 Sulfonated poly(aryletherketone) membranes

The poly(arylether ketones) are a class of non-fluorinated polymers consisting of sequences of ether and carbonyl linkages between phenyl rings, that can either "ether-rich" like PEEK and PEEKK, or "ketone-rich" like PEK and PEKEKK. The most common material is polyetheretherketone (PEEK) which is commercially available under the name Victrex™ PEEK from ICI Advanced Materials. A number of groups are developing proton conducting polymer materials based on this classification of materials including ICI Victrex, Fuma-Tech and Axiva/Aventis/Hoechst. Sulfonated-PEEK (SPEEK) membranes were prepared as proton conductors in PEMFCs by Schneller et al. (1993). Sulfonation of polyetherketones can be carried out directly in concentrated sulfonic acid or oleum - the extent of sulfonation being controlled by the reaction time and temperature (Bailly *et al.*, 1987; B. Bauer *et al.*, 1994, 1995). Direct sulfonation of PEEK can give materials with a wide range of equivalent weights to form SPEEK. However, the complete sulfonation of the polymer results in a fully water-soluble product. A sulfonation level of around 60% was found to be a good compromise between the conductivity and mechanical properties of membranes. The backbone of SPEEK is less hydrophobic than the backbone of Nafion®, and the sulfonic acid functional group is less acidic (Kreuer, 2001). Various studies have been made on the conductivity of SPEEK (Alberti *et al.*, 2001; B. Bauer *et al.*, 2000; Kobayashi *et al.*, 1998; Kreuer, 1997, 2001; Linkous *et al.*, 1998; P. Xing *et al.*, 2004; Zaidi *et al.*, 2000). The conductivity increases as a function of the degree of sulfonation, the ambient relative humidity, temperature and thermal history. The conductivity of these materials was found to be high at room temperature (Soczka-Guth *et al.*, 1999). In SPEEK with 65% sites sulfonated, the conductivity was higher than that of Nafion® 117 measured under the same conditions - the conductivity reaching 4×10^{-2} S/cm at 100°C and 100% RH (Linkous *et al.*, 1998). SPEEK membranes exhibit at 160°C and 75% RH, sufficiently high values of protonic conductivity - typically $5 \times 10^{-2} - 6 \times 10^{-2}$ S/cm - for possible applications in low temperature fuel cells (Alberti *et al.*, 2001). The dependence of the conductivity on RH is more marked for SPEEK than for Nafion® under the same conditions (B. Bauer *et al.*, 2000). Sulfonated polyaryls have been demonstrated to suffer from hydroxyl radical initiated degradation (Hubner & Roduner, 1999). In contrast, SPEEK was found to be durable under fuel cell conditions over several thousand hours by Kreuer (2001). The brittleness of SPEEK makes their handling difficult and may lead to mechanical membrane failure during operation. These types of membranes become very brittle when drying out. SPEEK can also be chemically cross-linked to reduce membrane swelling and increase its mechanical strength. Materials prepared by cross-linking are comparable to commercial Nafion® in terms of their mechanical strength and proton conductivity (Yen *et al.*, 1998). Kerres and co-workers prepared novel acid-base polymer blend membranes composed of SPEEK as the acidic compound, and of P4VP or PBI as the basic compounds (Kerres *et al.*, 1999; Jörissen *et al.*, 2002).

6. Organic/inorganic nanocomposite membranes

Membrane electrode assembly (MEA) is the basic component of the single cell of a stack. The proton exchange membrane (PEM) is the key element of this component, which separates the electrode structure to prevent the mixing of reactant gases and the formation of an electrical short. This makes its properties, functionality, cost and reliability very important for real cell operations. Up to now perfluorinated sulfonic acid (PFSA) membranes have been the best choice for commercial low temperature polymer products (<80°C). The advantages of PFSA membranes are:

i.   Their strong stability in oxidative and reduction media due to the structure of the polytetrafluoroethylene backbone; and

ii. Their proton conductivity, which can be as high as 0.2 S cm^{-1} in full hydrated polymer electrolyte fuel cell.

When used at elevated temperatures, however, PEMFC performances decrease. This decrease is related to: dehydration of the membrane; reduction of ionic conductivity;

decreases in affinity with water; loss of mechanical strength through a softening of the polymer backbone; and parasitic losses (the high level of gas permeation). There are several reasons for the development of high temperature membranes (Savadogo, 2004):

i.   The operation of PEMFC at temperature above 140°C is receiving world-wide attention because fuel selection remains straightforward, and a number of fuels, including reformed hydrogen with high CO content and light hydrocarbons (alcohol, natural gas, propane, etc.) are still being considered for PEMFC application. Accordingly, cell temperature operation at temperatures above 140°C is of great interest because, in this temperature range, anode catalyst poisoning by CO is less important and the kinetics of fuel oxidation will be improved and the efficiency of the cell significantly enhanced. High temperature cell operation will contribute to reducing the complexity of the hydrocarbon fuel cell system. Some other advantages of operating PEMFC at high temperature are: a reduction in the use of expensive catalysts; and minimization of the problems related to electrode flooding. Light hydrocarbons may be potential energy vectors for PEMFC, which may lead to the development of suitable membranes that are stable in high temperature operating conditions and prevent fuel crossover.

ii. Enhancement of gas transport in the electrode layers is also expected because no liquid water will be present in the cell at these temperatures. Membrane proton conductivity should be dependent on water content at these temperatures; consequently, it is not necessary to humidify the gas before it enters the stack. This may help improve the kinetics of mass transport and simplify the fuel cell system. In particular, the kinetics of oxygen reduction reaction could be improved, by at least three orders of magnitude, if we increase the operating temperature from 25 to 130°C. PFSA membranes cannot be used in PEMFC operating above temperatures around 100°C, because at these temperatures they will lose their mechanical properties and their swelling properties will be lowered. They do not perform well above 90°C in a hydrocarbon PEMFC and above 85°C in hydrogen PEMFC. The boiling point of water can be raised by increasing the operating pressure above 3 bar, which may correspond to a boiling point of water of about 135°C. But raising the pressure of PEMFC is undesirable from an efficiency point of view.

One of the main drawbacks of DMFC (direct methanol fuel cell) is the slow methanol oxidation kinetics. An increase in the operating temperature of the DMFC from 90 to about 140°C is highly desirable. Also operation at high temperature will enhanced CO tolerant when a reformate hydrogen is used in H_2-PEMFC. One approach to achieve water retention at high temperature is to fabricate a composite membrane constituted of organic proton conductor and inorganic materials. The organic/inorganic composite proton conductors are developed to overcome the breakdown of the actual state-of-the-art membranes (i.e. PFSA membranes: Nafion® (DuPont), Dow, Flemion® (Asahi Glass Corporation) and Aciplex® (Asahi Chemicals)). Thus, increasing the operating temperature above 100°C, reduced methanol permeability (methanol crossover), increasing the water retention and also increasing the mechanical and thermal stability of the composite membranes.

The method of inclusion of inorganic proton conductor or inorganic particle has involved a bulk powder dispersed in a polymer solution, leading specifically to particles of highly dispersed inorganic fillers of particle size in the sub-micronic range. These methods make use of mild chemistry technique, including intercalation/exfoliation, sol-gel chemistry, and ion-exchange (Bonnet *et al.*, 2000; Jones & Rozière, 2001). Such approach generally avoid any sedimentation of the inorganic component, intimacy of contact between the inorganic and organic components at the molecular level assures the greatest possible interface and, at such small particle size, the mechanical properties can be improved compared with those of a polymer-only membrane (Jones & Rozière, 2001, 2003). In addition, since in many proton conductors of conductivity suitable for electrochemical applications the proton transfer process takes place on the surface of the particles, increase in surface area (small particle size) will increase the conductivity (Jones & Rozière, 2001).

This concept was suggested by Watanabe *et al.* (1995, 1996) and is based on the development of self-humidifying composite membranes. The membranes are fabricated from the dispersion of nano-particles of Pt in a thin Nafion® film ($\approx$ 50 μm). Membranes fabricated based on this concept should not require external humidification and should suppress the crossover of H_2 and O_2. The dispersed particles should catalyze the oxidation and the reduction of the crossover H_2 and O_2 respectively. Water from this reaction is directly used to humidify the membrane. This is supposed to result in a more stable operation of the cell at 80°C without any external humidification of the membrane (Watanabe *et al.*, 1998).

6.1 Organic/silica nanocomposite membranes

Silica as an additive to Nafion® was widely studies. Both recast Nafion® (Adjemian *et al.*, 2002a,2002b; Antonucci *et al.*, 1999; Arimura *et al.*, 1999; Dimitrova *et al.*, 2002a,2002b) and Nafion® film (e.g. Nafion® 117) (Adjemian *et al.*, 2002b; Baradie *et al.*, 2000; Jung *et al.*, 2002) are used in the fabrication of the composite membrane. Organic–silica composite membranes have been prepared according to several methods by casting mixtures such as: using silicon dioxide particles (e.g. Aerosil A380 from Degussa) (Antonucci *et al.*, 1999; Aricò *et al.*, 1998; Arimura *et al.*, 1999; Dimitrova *et al.*, 2002a,2002b), diphenylsilicate (DPS) (Liang *et al.*, 2006), the other one is to introduce silica oxide incorporated via *in situ* sol-gel reaction of tetraethoxysilane (TEOS) (Adjemian *et al.*, 2002b; Baradie *et al.*, 2000; Deng *et al.*, 1998; R.C. R.C. Jiang *et al.*, 2006b; Jung *et al.*, 2002; Mauritz *et al.*, 1995, 1998). Nafion®–silica membranes shows good performance at T > 100°C due to low levels of dehydration. Nafion®–silica membranes were prepared by mixing Nafion® ionomer (5 wt.%) with 3 wt.% SiO_2 followed by a regular membrane casting procedure. In the final stage, the membranes were heat treated at 160°C for 10 min to achieve both a high crystallinity and high mechanical stability (Antonucci *et al.*, 1999). A DMFC utilizing these membranes was tested under galvanostatic conditions at 500 mA cm^{-2}. The voltage initially decreased from 0.42 to 0.36 V but then remained stable for 8 h. The performance decrease is due to adsorption of poisoning species, which appears to be a reversible process at 145°C (removed by short circuit discharging in the presence of water). These membranes demonstrated higher performance with increasing temperature. A nano-particles possessed a core-shell structure consisting of silica core (< 10 nm) and a densely grafted oligometric ionomer layer was incorporated into Nafion® matrix to form a composite membrane. The polyelectrolyte-grafted silica particles [P(SPA)-SiO$_2$] was dispersed in Nafion® solution and a composite membrane was formed by recasting process. The proton conductivity of Nafion® membrane containing P(SPA)-SiO$_2$ (4 wt.%) is significantly higher than that of unmodified recast Nafion® in the range 25 to 80°C. The composite membrane offers superior cell performance over unmodified recast Nafion® at both operating temperature of 50 and 80°C. At 50°C, the maximum power output of the composite membrane is about 1.8 times greater than of the Nafion® membrane and at 80°C, the ratio becomes 1.5 (Tay *et al.*, 2008).

Adjemian *et al.* (2002b) prepared composite membranes by either impregnating an extruded film via sol-gel processing of tetraethoxysilane (TEOS), or by preparing a recast film, using solubilized PFSA and silicon oxide polymer/gel. TEOS when reacted with water in an acidic medium undergoes polymerization to form a mixture of silicas (SiO_x) and siloxane polymer with terminal hydroxide and ethoxide groups(SiO_x/-OH/-OEt). When PFSAs are used as the acidic medium, the SiO_x/siloxane polymer forms within the membrane. Composite membranes were tested in fuel cell operating with pre-humidified reactant gases at temperature of 130°C and a pressure of 3 atm. The PFSA/silicon oxide composite membranes shows resistivities 50% lower than their respective unmodified PFSA under the same operating conditions. The observed resistivity trend from best to worst is as follows: Aciplex® 1004/silicon oxide > Nafion® 112/silicon oxide > Nafion® 105/silicon oxide > Aciplex® recast/silicon oxide > Nafion® recast/silicon oxide > Nafion® 115/silicon oxide.

Recently a new approach to make composite membrane was introduced, where a functionalized silica is used as a filler to make the composite membranes (Li *et al.*, 2006; Y.F. Lin *et al.*, 2007; Sambandam & Ramani, 2007 ; Su *et al.*, 2007; Tung & Hwang, 2007). Sol-gel derived sulfonated diphenyldimethoxysilane (SDDS) with hydrophilic –SO_3H functional groups were used as the additive to reduce the methanol permeability of Nafion® (Li *et al.*, 2006). The Nafion®-SDDS nanocomposite membranes were prepared by mixing Nafion® – dimethyl formamide (DMF) solutions with SDDS sol and casting to membranes. The proton conductivity of composite membrane decreased compared with commercial Nafion® membranes. This is partly because (i) the relative low conductivity of organosilica, (ii) the slightly tortuous path through the membrane which is caused by the embedding of the organosilica into the hydrophilic clusters, (iii) and the hydrophobic phenyl groups of the organosilica which change the distribution of the hydrophilic/hydrophobic phases therefore reduce the water content of the membrane. The proton conductivity decreases with the increase of the fillers. On the other hand, the methanol permeability is reduced with the increase of the SDDS content. The methanol permeability drops by a factor of 0.41, 0.61, 0.67 and 0.71 times for nanocomposite with loading of 5, 10, 20 and 25 wt.%, respectively as compared to bare recast Nafion® (Li *et al.*, 2006).

Sulfonic acid functionalized silica was synthesized by condensation of MPTMS (3-mercaptopropyltrimethoxy silane) precursor through a sol-gel approach. Sulfonated poly(ether ether ketone) composite with sulfonic acid functionalized silica were prepared by casting (Sambandam & Ramani, 2007). At 80°C and 75% RH (relative humidity) the measured conductivity was 0.05 S cm[-1] for SPEEK containing 10% sulfonic acid functionalized silica and 0.02 S cm[-1] for the plain SPEEK membrane. At 80°C and 50% RH the measured conductivity was 0.018 S cm[-1] for SPEEK containing 10% sulfonic acid functionalized silica and 0.004 S cm[-1] for the plain SPEEK membrane.

L64 copolymer-templated mesoporous SiO_2, functionalized with perfluoroalkylsulfonic acid was prepared (Y.F. Lin *et al.*, 2007). A condensation reaction between the surface silanol groups of the mesoporous silicas and 1,2,2-trifluoro-2-hydroxy-1-trifluoromethylethane sulfonic acid Beta-sultone was conducted. Nafion®/functionalized mesoporous silica composite membranes were prepared via homogeneous dispersive mixing and the solvent casting method. The room temperature proton conductivity of full hydration composite membrane was increased from 0.10 to 0.12 S cm^{-1} as the M-SiO$_2$-SO$_3$H content increased from 0 to 3 wt.%. Methanol permeability decreases with increasing the content of M-SiO$_2$-SO$_3$H, where methanol permeability was 4.5 × 10^{-6} cm^2 s^{-1}, which was 30% lower than unmodified Nafion®. The current densities measured with composite membranes containing 0, 1, 3 and 5 wt.% M-SiO$_2$-SO$_3$H, were 51, 66, 80 and 70 mA cm^{-2}, respectively, at a potential of 0.2 V. Moreover, all composite membranes containing M-SiO$_2$-SO$_3$H performed better at high current density than did unmodified Nafion®.

Nanocomposite proton exchange membranes were prepared from sulfonated poly(phtalazinone ether ketone) (SPPEK) and various amounts of sulfonated silica nanoparticles (silica-SO$_3$H) (Su et al., 2007). The use of silica-SO$_3$H compensates for the decrease in ion exchange capacity of membranes observed when no-sulfonated nano-fillers are utilized. The strong –SO$_3$H/-SO$_3$H interaction between SPPEK chains and silica-SO$_3$H particles leads to ionic cross-linking in the membrane structure, which increases both the thermal stability and methanol resistance of the membranes. The membrane with 7.5 phr of silica-SO$_3$H (phr = g of silica-SO$_3$H / 100 g of SPPEK in membranes) exhibits low methanol crossover, high bound-water content, and a proton conductivity of 3.6 fold increase to that of the unmodified SPPEK membrane. Nafion®/hydrated phosphor-silicate composite membrane was synthesis by Tung and Hwang (2007). The phosphor-silicate glass, with a nominal composition of 30% P$_2$O$_5$ and 70% SiO$_2$ (molar ratio) (called 30P70Si), was prepared by the accelerated sol-gel process, where tetraethylorthosolicate and trimethyl phosphate are used as precursors. It was found that the methanol permeability decreases dramatically with increased SiO$_2$-P$_2$O$_5$ content and the proton conductivity only decreases slightly , as a consequence the selectivity of the hybrid membranes are higher than unmodified Nafion® membrane.

6.2 Organic/heteropolyacid (HPA) nanocomposite membranes

Perfluorosulfonic acid based organic/ inorganic composite membranes with different heteropolyacid (HPA) additives have been investigated as alternate materials for low humidity PEMFC operation (Giordano et al., 1996; Ramani et al., 2004, 2005a,2005b; Tazi & Savadogo, 2000, 2001). Two major factors limiting the performance of Nafion®/HPA composite membranes are (Ramani et al., 2005b): (i) the high solubility of the HPA additive and (ii) the large particle size of the inorganic additive within the membrane matrix (Ramani et al., 2004, 2005a). Stabilization technique have been developed (Ramani et al., 2005a) to limit the solubility of the HPA additive. Recast Nafion® with phosphotungstic acid (PTA) as HPA fillers were prepared by Ramani et al. (2005b). Three types of fillers were used PTA with 30-50 nm particle size, PTA with 1-2 μm particle size and TiO$_2$ with 1-2 μm particle size. The composite membranes had hydrogen crossover currents on the order of 1-5 mA cm^{-2}, with the crossover flux decreasing and approaching the value for recast Nafion® as the particle size was reduced. A 25 μm thick composite membrane with PTA (1-2 μm particle size) had an area-specific resistance of 0.22 Ω cm^{-2} at 120°C and 35% RH, while the corresponding value for a 25 μm thick composite membrane with PTA (30-50 nm) was 0.16 Ω cm^{-2}. The latter membrane compared favorably with recast Nafion®, which had an area-specific resistance of 0.19 Ω cm^{-2} under the same conditions. Savadogo and co-workers (Savadogo, 2004; Tazi & Savadogo, 2000, 2001; Tian & Savadogo, 2005) prepared composite membranes constituted of recast Nafion®and mixed with appropriate concentration of HPA, namely, silicotungstic acid (STA), phosphotungstic acid (PTA) and phosphomolybdic acid (PMA). It was shown that the water uptake of the various membranes increases in this order: Nafion®117 (27%) < Nafion®/STA (60%) < Nafion®/PTA (70%) < Nafion®/PMA (95%). The ionic conductivity increases in the order Nafion®117 (1.3×10^{-2} S cm^{-1})) < Nafion®/PMA (1.5×10^{-2} S cm^{-1}) < Nafion®/PTA (2.5×10^{-2} S cm^{-1}) < Nafion®/STA (9.5×10^{-2} S cm^{-1}). The tensile strength of the membranes decreases in the order: Nafion®117 (15000 Pa) < Nafion®/STA (14000 Pa) < Nafion®/PMA (8000 Pa) < Nafion®/PTA (3000 Pa), while their deformation (ε_{max}) changes in the order : Nafion®/STA (45%) < Nafion®/PMA (70%) < Nafion®/PTA (170%) < Nafion®117 (384%). The current density at 0.600 V of the PEMFCs based on the various membranes varies in the order: Nafion®117 (640 mA cm^{-2}) < Nafion®/STA (695 mA cm^{-2}) < Nafion®/PTA (810 mA cm^{-2}) < Nafion®/PMA (940 mA cm^{-2}).

Tazi and Savadogo (2000) fabricated Nafion® membranes containing silicotungstic acid and thiophene. They reported an increase of up to 60% of water uptake and a considerable improvement in the fuel cell current density, when compared to the plain Nafion® membrane. Dimitrova et al. (2002a) prepared a recast Nafion®-based composite membrane containing molybdophosphoric acid. This composite membrane exhibit significantly higher conductivity in comparison to Nafion® 117 and pure recast Nafion®. An enhancement of a factor of 3 in the conductivity at 90°C was observed. Zaidi et al. (2000) prepared a series of composite membranes using SPEEK as polymer matrix and tungstophosphoric acid (TPA), its sodium salt (Na-TPA) and molybdophosphoric acid (MoPA) as inorganic fillers. The conductivity of the composite membranes exceeded 10^{-2} S/cm at room temperature and reached values of about 10^{-1} S/cm above 100°C. From the DSC (Differential Scanning Calorimeter) studies, it was indicated that the glass transition temperature of the SPEEK/HPA composite membrane increases due to the incorporation of solid HPA into SPEEK membrane. This increase in the glass transition temperature was attributed to an intermolecular interaction between SPEEK and HPA. Staiti et al. (2001) prepared Nafion®(recast)-silica composite membranes doped with phosphotungstic (PWA) and silicotungstic (SiWA) acids for application in direct methanol fuel cell at high temperature (145°C). The phosphotungstic acid-based membrane showed better electrochemical characteristics at high current densities with respect to both silicotungstic acid-modified membrane and silica- Nafion® membrane. The best electrochemical performance is obtained with the PWA-based membrane, which gives a maximum power density of 400 mW cm^{-2} at current density of about 1.4 A cm^{-2} under oxygen feed operation at 145°C. Maximum power density of 340 mW cm^{-2} is obtained from the fuel cell which uses the silica-modified membrane, whereas a lower performance was achieved with the SiWA-based membrane. The maximum power density obtained in air with the PWA-based membrane is 250 mW cm^{-2} at 145°C, and 210 mW cm^{-2} with the Nafion-SiO$_2$ membrane at the same temperature.

Shao et al. (2004) prepared Nafion®/silicon oxide (SiO$_2$)/phosphotungstic acid (PWA) and Nafion®/silicon oxide composite membranes for H$_2$/O$_2$ proton exchange membrane fuel cells operated above 100°C. It was found that the composite membranes showed a higher water uptake compared with the Nafion® recast membrane. The proton conductivity of the composite membranes appeared to be similar to that of the native Nafion® membrane at high temperatures and 100% relative humidity (RH), however, it was much higher at low RH. When the composite membranes viz. Nafion®/SiO$_2$/PWA and Nafion®/SiO$_2$ were employed as an electrolyte in H_2/O_2 PEMFC, the higher current density values (540 and 320 mA cm^{-2} at 0.4 V, respectively) were obtained than that of the Nafion®115 membrane (95 mA cm^{-2}), under the operating conditions of 110°C and 70% RH. A similar membrane was prepared by Aricò et al. (2003a,2003b). Sulfonic-functionalized heteropolyacid-SiO_2 nanoparticles were synthesized by grafting and oxidizing of a thiol-silane compound onto the heteropolyacid-SiO_2 nanoparticle surface (H.J. Kim et al., 2006). The composite membrane containing the sulfonic-functionalized heteropolyacid-SiO_2 nanoparticles was prepared by blending with Nafion® ionomer. TG-DTA analysis showed that the composite membrane was thermally stable up to 290°C. The DMFC performance of the composite membrane increased the operating temperature from 80 to 200°C. The function of the sulfonic-functionalized heteropolyacid-SiO_2 nanoparticles was to provide a proton carrier and act as a water reservoir in the composite membrane at elevated temperature. The power density was 33 mW cm^{-2} at 80°C, 39 mW cm^{-2} at 160°C, 44 mW cm^{-2} at 200°C, respectively.

SPEEK-silica membranes doped with phosphotungstic acid (PWA) was synthesized by Colicchio et al. (2009). The silica is generated insitu via the water free sol-gel process of polyethoxysiloxane (PEOS), a liquid hyperbranched inorganic polymer of low viscosity. PEOS was used as silica precursor instead of the corresponding monomeric tetraethoxysilane (TEOS). At 100°C and 90% RH the membrane prepared with PEOS (silica content = 20 wt.%) shows two times higher conductivity than the pure SPEEK. The addition of small amount of PWA (2 wt.% of the total solid content) introduce in the early stage of membrane preparation brings to a further increase in conductivity (more than three times the pure SPEEK). Different classes of composite membranes containing HPA and silica were developed, namely, phosphomolybdic acid (PMA)/phosphotungstic acid (PWA)- P_2O_5-SiO_2 glass electrolyte (Uma & Nogami, 2007), poly(vinyl alcohol) (PVA)/sulfosuccinic acid (SSA)/silica hybrid membrane (D.S. Kim et al., 2004), PVA/SiO_2/ SiW (silicotungstic acid) (Shanmugam et al., 2006), PVA/PWA/SiO_2 (W. Xu et al., 2004), polyethylene oxide (PEO)/PWA/SiO_2 (Honma et al., 2002), polyethylene glycol (PEG)/4-dodecylbenzene sulfonic acid (DBSA)/SiO_2 (H.Y. Chang et al., 2003), PWA-doped PEG/SiO_2 (C.W. Lin et al., 2005).

6.3 Organic/TiO₂ nanocomposite membranes

Nanosized titanium oxide was synthesized by sol-gel hydrolyzing an alcoholic solution of $Ti(OiPr)_4$ by Baglio et al. (2005). Thermal treatments at different temperature, namely 500, 650 and 800°C, were performed to tailor the oxide powder properties. The crystallite size for the three sample was found to be 12, 22 and 39 nm, respectively. A composite membrane Nafion®/ 5 wt.% TiO_2 was prepared by using the recast procedure. The composite membrane thickness was about 100 µm. A maximum power density of 350 mW cm^{-2} was obtained at 145°C with the cell equipped with the composite membrane containing TiO_2 calcined at 500°C. Sacca et al. (2005) synthesized TiO_2 powder by the sol-gel method starting with a $Ti(OiPr)_4$ and calcined a 400°C. This powder was made of spherical particles with a grain size between 5 and 20 nm. A recast Nafion® with 3 wt.% TiO_2 composite membrane was prepared, the thickness of the membrane was 100 µm. The proton conductivity of different membranes were measured at two different values of relative humidity (RH), 100 and 85% RH, respectively, and simulating the cell operating conditions in the temperature range from 80 to 130°C. Nafion® recast (70 µm thickness) has the lower conductivity ranging from 0.12 to 0.14 S cm^{-1}, while the composite Nafion®/TiO$_2$ showed highest value than Nafion® 115 (125 μm) were value in the range 0.15-0.18 S cm^{-1}. A power density of 0.514 W cm^{-2} for Nafion® / 3 wt.% TiO$_2$ composite against 0.354 W cm^{-2} for Nafion® 115 at 0.56 V and at T = 110°C was recorded. At 120°C, Nafion® 115 was damaged while the composite Nafion®/TiO$_2$ membrane continued to work up to 130°C by reaching a power density of about 0.254 W cm^{-2} at 0.5 V.

Hybride membranes based on highly sulfonated poly(ether ether ketone) (SPEEK, DS = 0.9) where titania network was dispersed by *insitu* sol-gel reactions were prepared by Di Vona *et al.* (2007). Titania network was introduced following two routes: route 1 using titanium tetrabutoxide (Ti(OBu)$_4$) and pyridine, and route 2 uses Ti(OBu)$_4$ and 2,4-pentandione. Composite membranes prepared by route 2 showed a good conductivity property that can be attributed to the structural characteristics of the inorganic network generated in the presence of a chelating agent. This membrane shows a stable value (σ = 5.8 × 10^{-2} S cm^{-1}) at 120°C in fully hydrated conditions. Jian-hua *et al.* (2008) prepared a composite Nafion®/TiO$_2$ membranes by carrying out *insitu* sol-gel reaction of Ti (OC$_4$H$_9$)$_4$ followed by hydrolyzation-condensation in Nafion® 112, 1135 and 115. TiO$_2$ prepared with this method was found to be 4 nm. TiO$_2$ contents were 1.23, 2.47 and 3.16 wt.% for Nafion® 112/TiO$_2$, Nafion® 1135/TiO$_2$ and Nafion® 115/TiO$_2$, respectively. The polarization characteristics of all three MEAs with the membranes containing TiO$_2$ were improved significantly comparing with those of pure Nafion® film. A mixture of titanium tetraisopropoxide (TTIP) and PEG 1000 were used to prepare the titania sol by Liu *et al.* (2006). The average particle size of 20 nm was reported. The formed sol was deposited on the surface of Nafion® 112 membranes by spin coating. The TiO$_2$ film is dense and well attached to the membrane, but some cracks in the membrane coated with diluted titania sol (e.g. 0.002 mg cm^{-2}), while the membrane coated with thick titania sol (e.g. 0.021 mg cm^{-2}) are very dense and cracks free. The proton conductivity of nano-TiO$_2$-coated Nafion® membranes at 25 and 80°C were recorded with different TiO$_2$ content. It was found that the maximum conductivity was with uncoated Nafion® 112, with values of 0.027 and 0.041 S cm^{-1} for 25 and 80°C, respectively. The conductivity of coated Nafion® decreases with increasing titania content. On the other hand, methanol permeability of the coated membranes decreases with increasing TiO$_2$ content, namely from 3.2 × 10^{-6} to 1.7 × 10^{-6} cm^2 s^{-1} at 25°C, and from 12.5 × 10^{-6} to 4.6 × 10^{-6} cm^2 s^{-1} at 85°C. Thus the rise in temperature leads to a strong increase in permeation by a factor of about 3. The methanol permeability of the unmodified Nafion® 112 membrane was found to be 3.6 × 10^{-6} and 13 × 10^{-6} cm^2 s^{-1} at 25 and 85°C, respectively. The cell performance with titania coated membrane with a content of 0.009 mg cm^{-2} exhibits a higher voltage than cells with Nafion® membrane or with the other coated membranes. Nafion® 112 delivered a maximum power density of 37 mV cm^{-2} at a current density of 200 mA cm^{-2}. A maximum power density of 44 mW cm^{-2} is obtained from a fuel cell that employs the titania-coated membrane with 0.009 mg cm^{-2} content.

6.4 Organic/zirconia and sulfated zirconia nanocomposite membranes

Zirconia as an inorganic filler was added to polymeric proton conductor membranes (Aricò *et al.*, 2003b, 2004; Nunes *et al.*, 2002; Sacca *et al.*, 2006; Silva *et al.*, 2005a, 2005b). The incorporation of zirconia should increase the working temperature, water retention and mechanical stability of the composite membrane. Organic / inorganic composite membranes based on SPEK and SPEEK for application in direct methanol fuel cell were synthesized by Nunes and co-workers (Nunes *et al.*, 2002). The inorganic fillers were introduced via *in situ* generation of SiO_2, TiO_2 or ZrO_2. The modification with ZrO_2 led to a 60-fold reduction of the methanol flux. However, a 13-fold reduction of conductivity was also observed.

Recast Nafion® composite membranes containing three different percentages (5%, 10% and 20%, w/w) of commercial ZrO_2 as an inorganic filler were tested in fuel cell in a temperature range of 80-130°C in humidified H_2/air gases at 3.0 bar abs by Sacca *et al.* (2006). The introduction of 5 wt.% ZrO_2 in Nafion® produces no evidence changes in the cell performance, while a better performance with 10 wt.% ZrO_2 in Nafion® was obtained with a power density greater than 600 mW cm^{-2} at 0.6 V both at 80°C and 110°C. The good performance of 10 wt.% ZrO_2 in Nafion® was maintained at 130°C with gas humidification of 85% RH, with a maximum power density of about 400 mW cm^{-2} was obtained in the potential range of 0.5-0.6 V. Silva *et al.* (2005a,2005b) prepared SPEEK/ZrO_2 composite membranes using *insitu* formation of zirconia with zirconium tetrapropylate as alkoxide and acetyl acetone as chelating agent. The water/alkoxide ratio was always maintained higher than 1 to ensure the formation of a finely dispersed inorganic phase in the polymer solution. The thickness of the prepared membranes with 0.0, 2.5, 5.0, 7.5, 10.0, 12.5 wt.% of zirconium oxide were 188, 175, 133, 146, 128, 106 μm, respectively. The proton conductivity of the composite membranes was measured at 25°C and it was found that it decreases continuously with the ZrO_2 content. Pervaporation experiments at 55°C showed that the membrane permeability towards methanol decreases with the amount of ZrO_2. Composite SPEEK with 5.0, 7.5, 10.0 wt.% of ZrO_2 were tested in fuel cell. The performance of 12.5 wt.% ZrO_2 composite was very low due to the high ohmic resistance of the corresponding MEA. The membrane 5 wt.% ZrO_2 presents the best DMFC performance among all the studied MEAs. Three types of superacidic sulfated zirconia (S-ZrO_2) were prepared by different methods using hydrated zirconia and sulfuric acid by Hara and Miyayama (2004). Their proton conductivities were evaluated at 20-150°C under saturated water vapor pressure. It was found that the concentration on S-ZrO_2 varied largely depending on the method of preparation. The S/Zr atomic ratio changed from 0.046 for the sample prepared through a mixture of hydrated zirconia powder and sulfuric acid to 0.35 for sample prepared through a mixture of hydrated zirconia sol and sulfuric acid. A powder compact of the former S-ZrO_2 showed a proton conductivity of 4 × 10^{-2} S cm^{-1} at 70°C and 8 × 10^{-3} S cm^{-1} at 150°C, whereas that of the latter S-ZrO_2 exhibited a high conductivity of 5 × 10^{-2} S cm^{-1} at 60-150°C.

S. Ren *et al.* (2006) prepared sulfated zirconia/ Nafion® 115 nanocomposite membrane by ion exchange of zirconium ions into the Nafion® followed by precipitation of sulphated ZrO_2 by treatment in H_2SO_4. The incorporation of sulfated zirconia increases water uptake by the Nafion® membrane, and more water is absorbed than an unmodified membrane at high temperatures. The membrane proton conductivity is decreased slightly by ZrO_2 impregnation. The proton conductivity of Nafion® 115 membrane was found to be 1.5 × 10^{-2} S cm^{-1} at 25°C, while that of S-ZrO_2/ Nafion® 115 membrane is decreased to 5.0 × 10^{-3} S cm^{-1} at 25°C. At 110°C and above, the proton conductivity of S-ZrO_2/ Nafion® 115 membrane is more than one-half that of the Nafion® 115 membrane. Fine particle superacidic sulfated zirconia (S-ZrO_2) was synthesized by ameliorated method, and composite membranes with different S-ZrO_2 contents were prepared by a recasting procedure from a suspension of S-

ZrO_2 powder and Nafion® solution (Zhai *et al.*, 2006). The results showed that the IEC (Ion Exchange Capacity) of composite membrane increased with the content of S-ZrO_2 and S-ZrO_2 was found to be compatible with the Nafion® matrix. The incorporation of the S-ZrO_2 increased the crystallinity and also improved the initial degradation temperature of the composite membrane. The performance of single cell was the best when the S-ZrO_2 content was 15 wt.% and achieved 1.35 W cm^{-2} at 80°C and 0.99 W cm^{-2} at 120°C based on H_2/O_2 and at a pressure of 2 atm, the performance of the single cell with optimized S-ZrO_2 was far more than that of the Nafion® at the same condition (e.g. 1.28 W cm^{-2} at 80°C and 0.75 W cm^{-2} at 120°C). A self-humidifying composite membrane based on Nafion® hybrid with SiO_2 supported sulfated zirconia particles (SiO_2-SZ) was fabricated and investigated for fuel cell application by Bi *et al.* (2008). The bi-functional SiO_2-SZ particles, possessing hygroscopic property and high proton conductivity, was incorporated in recast Nafion® membrane. The proton conductivity of Nafion®/SiO_2-SZ, Nafion®/SiO_2 and recast Nafion® under dry and wet H_2/O_2 conditions at 60°C were compared. The two composite membranes showed higher proton conductivity in contrast to the recast Nafion® membrane under 0% RH mode with the order Nafion®/SiO_2-SZ > Nafion®/SiO_2 > recast Nafion®. Under 100% RH mode, the Nafion®/SiO_2-SZ composite membrane also exhibited the highest proton conductivity values among the three membranes. The proton conductivity of Nafion®/SiO_2-SZ membrane was 6.95×10^{-2} S cm^{-1} and the value was higher than Nafion®/ SiO_2 membrane of 5.54×10^{-2} S cm^{-1} and recast Nafion® membrane of 6.55×10^{-2} S cm^{-1}. Single cell performance of these composite membranes were tested with wet H_2 and O_2 at 60°C. Nafion®/ SiO_2 composite membrane exhibited the worst output performance (0.864 W cm^{-2}) due to the increased proton conductive resistance caused by incorporated less proton conductivity of SiO_2 particles. In contrast, Nafion®/SiO_2-SZ composite membrane showed similar cell performance to recast Nafion® (1.045 W cm^{-2} vs. 1.014 W cm^{-2}). However, the single cell performance of Nafion®/SiO_2-SZ and Nafion®/SiO_2 membranes with dry H_2 and O_2 at 60°C were 0.980 and 0.742 W cm^{-2}, respectively. These results shows that the composite membrane perform better than unmodified Nafion® (i.e 0.635 W cm^{-2}) under dry condition and the composite membranes manifested a good water retention.

6.5 Organic/zirconium phosphate nanocomposite membranes

Layered zirconium phosphate (ZrP) and phosphonates were used as inorganic fillers of proton conducting polymeric membranes because they are proton conductors with good chemical and thermal stability. Under the most favourable conditions, their conductivity is around 10^{-2} S cm^{-1} for high surface ZrP (Alberti *et al.*, 1978; F. Bauer & Willert-Porada, 2005) and 10^{-1} S cm^{-1} for zirconium phosphate sulfophenylenphosphonates (Alberti *et al.*, 2004,2005a). ZrP can be added to Nafion® (Alberti *et al.*, 2007; F. Bauer & Willert-Porada, 2004,2005,2006a; Casiola *et al.*, 2008; Costamagna *et al.*, 2002; Grot & Rajendran, 1999; Helen *et al.*, 2006,2007; Hou *et al.*, 2008; R. Jiang *et al.*, 2006a; Kuan *et al.*, 2006; H.K. Lee *et al.*, 2004; Mitov *et al.*, 2006; Yang *et al.*, 2001a,2001b,2004), SPEEK (Bonnet *et al.*, 2000; Nunes *et al.*, 2002; Silva *et al.*, 2005b; Tchicaya-Bouckary *et al.*, 2002; Triphathi *et al.*, 2007,2009), SPEK (Nunes *et al.*, 2002; Ruffmann *et al.*, 2003) and difulfonated poly(arulene ether sulfone) (Hill *et al.*, 2006). A similar inorganic material derived from ZrP, named zirconium phosphate sulfophenylen-phophonate was also used as a filler with Nafion® (Y.T. Kim *et al.*, 2004), SPEEK (Krishnan *et al.*, 2006) and PVDF (polyvinyl -lidene fluoride) (Casiola *et al.*, 2005).

F. Bauer & Willert-Porada (2004,2005,2006a) impregnated Nafion® with different ZrP contents. The proton conductivity of unmodified Nafion® 117 and Nafion® 117/ZrP (21 wt.%) was measured at three different temperatures, 80, 100 and 130°C. It was found that the presence of ZrP decreased the proton conductivity in all cases. At high humidity the conductivity first increased from 80 to 100°C and decreased at 130°. The conductivity decrease is more pronounced in case of the unmodified Nafion®. DMFC performance was conducted with Nafion® 117 and Nafion® 117/ZrP composite membrane at 130°C and 4.6 bar at the anode and cathode. It was found that the power output of Nafion® was higher than that of the composite membranes (Nafion® 117/13 wt.% ZrP and Nafion® 117/26 wt.% ZrP). At 0.2 A cm^{-2}, a values of 420, 370 and 370 mV were measured for Nafion® 117, Nafion® 117/13 wt.% ZrP and Nafion® 117/26 wt.% ZrP, respectively. The crossover current was reduced by a factor of two as compared to the unmodified Nafion®. Also the two composite membranes tested exhibited a higher OCV than unmodified Nafion®, which also indicates lower methanol permeability. A values of 725, 768 and 760 mV were reported for Nafion® 117, Nafion® 117/13 wt.% ZrP and Nafion® 117/26 wt.% ZrP, respectively.

Yang and coworkers (Yang *et al.*, 2001a,2001b,2004; Costamagna *et al.*, 2002) introduced ZrP into Nafion® 115 through ion exchange of Zr^{4+} followed by precipitation of ZrP by treatment with phosphoric acid as described by Grot and Rajendran (1999). An MEA employing Nafion® 115/23 wt.% ZrP gave a H$_2$/O$_2$ PEMFC performance of about 1000 mA cm^{-2} at 0.45 V at a temperature of 130°C and a pressure of 3 bar, while unmodified Nafion® 115 gave 250 mA cm^{-2} at 0.45 V when operated under the same conditions of temperature and pressure. Similar experiment performed with recast Nafion® and recast Nafion®/36 wt.% ZrP composite confirmed an analogous improvement of performance of the composite membrane over the unmodified ones. The composite recast Nafion®/36 wt.% ZrP gave about 1500 mA cm^{-2} at 0.45 V at a temperature of 130°C and a pressure of 3 bar.

Alberti *et al.* (2005b,2007) prepared a recast Nafion® filled with ZrP according to the procedure described in the patent (Alberti *et al.*, 2005b). Zirconyl propionate was used instead of zirconyl oxychloride and the solutions were dissolved in DMF. The IEC (ion exchange capacity) of the prepared composite membrane was found to be higher than those previously reported for Nafion®/ZrP membranes prepared according to the exchange method (F. Bauer & Willert-Porada, 2006b; Yang *et al.*, 2004). The proton conductivity was found to decrease with increasing the filler loading, which is in agreement with the trend found for Nafion®/ZrP prepared by the exchange method (F. Bauer & Willert-Porada, 2005; Casiola *et al.*, 2008; Yang *et al.*, 2004). At constant RH, the logarithm of conductivity shows approximately the same linear dependence on ZrP loading in the RH range 50-90%. However, at 35% RH, the increase in the ZrP loading results in a larger conductivity decrease than that observed in the above RH range. A similar behavior was also reported for Nafion®/ZrP membranes obtained by the exchange method already at 50% RH (Yang *et al.*, 2004), thus confirming that the same type of filler prepared by using different procedures gives rise to different membrane properties. It was concluded that the main difference between pure Nafion® and composite membranes appear at low RH and high filler loading. It was reported that the Nafion® conductivity undergoes an irreversible decay above certain values of temperature and RH, which was attributed to an anisotropic swelling of the membrane, pressed between the electrodes, in the direction parallel to the membrane surface (Alberti *et al.*, 2001; Casiola *et al.*, 2006). It was also found that, at a given RH value, the decay temperature for composite Nafion®/ZrP membranes was higher than for pure recast Nafion® membranes prepared and thermally treated under the same conditions used for the composite sample. The conductivity of the pure Nafion® starts to decay at temperatures higher than 130°C, while the conductivity of the composite membrane is stable up to 140°C. Nafion® 115/23 wt.% ZrP was prepared by ion exchange and tested for DMFC by Hou *et al.* (2008). It was found that the liquid uptakes of Nafion® 115 and Nafion® 115/23 wt.% ZrP membranes increased linearly with increasing methanol concentration. The slope of the plot for Nafion® 115 was larger than for the composite membrane i.e. the liquid uptake of Nafion® 115 increased from 34.3% in 0M methanol solution to 58.6% in 10M methanol solution, while that of the composite membrane increased from 28.3% to 37.5% in the corresponding methanol solution. When 23 wt.% of ZrP was incorporated into Nafion® 115, the IEC of the resulting membrane increased significantly to 1.93 meq/g from a value of 0.909 meq/g for pure Nafion® 115. The proton conductivity at room temperature of Nafion® 115 and Nafion® 115/23 wt.% ZrP was found to be 0.10 and 0.084 S cm^{-1}, respectively. Also it was found that the methanol crossover through the composite membrane was suppressed. The DMFC test at 75°C and 5M methanol solution shows that the composite membrane performed better that the pure Nafion® 115, with a peak power density of 96.3 and 91.6 mW cm^{-2}, respectively. When the methanol concentration was further increase to 10M, the peak power density of DMFC with composite membrane was 76.19 mW cm^{-2}, which is higher than that for Nafion® (42.4 mW cm^{-2}). However, Bonnet *et al.* (2000) investigated the incorporation of ZrP in SPEEK. It was found that the conductivity of the composite membrane exceeded that of the polymer-only membrane, and increases with the amount of the filler (from 0-30 wt.%) up to 0.08 S cm^{-1} when measured at 100°C and 100% RH. A similar trend was also observed, when the RH varied from 50 to 100%. At all value of RH, the composite membrane SPEEK/20 wt.% ZrP conductivity was higher than that of non-modified SPEEK. A similar membrane was prepared by Tchicaya-Bouckary *et al.* (2002). The conductivity of SPEEK/25 wt.% ZrP was found to be weakly temperature dependence, the conductivity increases from 2×10^{-2} to 5×10^{-2} S cm^{-1} between 20 and 100°C at 100% RH. This composite membrane was tested in H$_2$/O$_2$ fuel cell at 100°C at an oxygen pressure of 3.6 bars absolute. A value of 1 A cm^{-2} at 0.6 V was reported. These results are much better than that reported for Nafion® 115/ ZrP (Costamagna *et al.*, 2002) which provided ca. 0.7 A cm^{-2} at 0.6 V, 130°C and 3 bars pressure. Nunes and co-workers studied the incorporation of ZrP in SPEEK and SPEK (Nunes *et al.*, 2002; Ruffmann *et al.*, 2003; Silva *et al.*, 2005c). ZrP was prepared according to the procedure described by Belyakov & Linkov (1999). It was reported that the incorporation of ZrP did not lead to a particular reduction of water and methanol permeability, and the proton conductivity at 25°C was decreased to the same extent (44 mS cm^{-1} for a SPEK/ 20 wt.% ZrP and 50 mS cm^{-1} for a pure SPEK membranes). A good values of proton conductivities were measured for membranes with 70/20/10 and 69/17/14 wt.% SPEK/ZrP/ZrO$_2$ where a conductivities of 45 and 35 mS cm^{-1} were measured, respectively (Nunes *et al.*, 2002). ZrP pretreated with *n*-propylamine and PBI was incorporated with SPEEK (Silva *et al.*, 2005c), the proton conductivity of the composite membranes decreases with the amount of inorganic incorporation. On the other hand, methanol and water permeability in the pervaporation experiments at 55°C showed that it decrease with the amount of inorganic incorporation. Similar trend was found for the composite membranes permeability towards nitrogen, oxygen and carbon dioxide. The SPEEK composite membranes were tested in a DMFC at 110°C, it was found that the unmodified SPEEK (SD = 42%) membrane presented the maximum power density output. It achieved an output power density value of 10.4 mW cm^{-1} for 51.8 mA cm^{-2}. The unmodified membrane with SD = 68% could not be characterized due to its instability (high swelling or even solubility). However, the SPEEK (SD=68%)/20 wt.% ZrP/11.2 wt.% PBI had even higher power density than the membrane with SD = 42% for current density lower than 25 mA cm^{-2}. When the relative humidity at the cathode feed was increased to 138%, the SPEEK (SD=68%)/20 wt.% ZrP/11.2 wt.% PBI membrane had the best performance, with an output power density value of 14.7 mW cm^{-1} for 58.8 mA cm^{-2} (Silva et al., 2005c). However, the filler addition to SPEEK (SD = 42%) besides reducing the crossover had an excessive (negative) effect on the proton conductivity.

Zirconium phosphate sulfophenylphosphate, a functionalized ZrP, was incorporated in Nafion® (Casiola et al., 2008; Y.T. Kim et al., 2004), SPEEK (Bonnet et al., 2000; Krishnan et al., 2006) and PVA (Casiola et al., 2005). ZrP sulfophenylphosphates (ZrSPP) are a class of layered materials exhibiting proton conductivity comparable with that of Nafion membranes (i.e. 0.07-0.1 S cm^{-1} at 100°C and 100% RH) due to the presence of the –SO$_3$H groups in the interlayer region (Alberti et al., 2005a). The functionalization of the ZrP nanoparticles with SPP is therefore expected to increase the conductivity of the Nafion®/ZrP membranes. These phosphonates are ideally obtained by partial replacement of SPP groups for the phosphate groups of ZrP (Casiola et al., 2008). Nafion® 117/ZrSPP composite membranes was found to have a higher conductivity than the parent Nafion® 117/ 20 wt.% ZrP and pure Nafion® 117 membrane at 100°C and RH between 30-90%, with highest value approaching 0.1 S cm^{-1} at RH = 90% (Casiola et al., 2008), while appreciable dehydration of Nafion® 117 resulted in drastic reduction of proton conductivity above 100°C (Y.T. Kim et al., 2004). However, the proton conductivity of Nafion®/12.5 wt.% ZrSPP composite membrane slightly increased up to 70°C and remained constant until 140°C, with a conductivity of 0.07 S cm^{-1} (Y.T. Kim et al., 2004).

6.6 Organic/palladium nanocomposite membranes

This approach is to utilize the unique properties of palladium which is permeable to protons, but very resistant to methanol transport. It was suggested first by Pu et al. (1995) where they used a palladium foil of 25 μm thick sandwiched between two Nafion® 115 sheets. They proved that with this approach methanol crossover can be reduced, but the cell performance will be lower due to the increase of the membrane thickness.

Choi et al. (2001) used the same approach by sputtering metallic palladium on the surface of a Nafion® 117 to plug the pores of Nafion®. The palladium film was found to be 20 nm. Methanol permeability was reduced from 2.392 × 10^{-6} cm^2 s^{-1} in unmodified Nafion® 117 to 1.7 × 10^{-6} cm^2 s^{-1} in Pd-sputtered membrane and the cell performance at 95°C was improved compared to the unmodified Nafion® 117 membrane. The methanol permeability reduction was confirmed by the high OCV obtained with the modified membranes. Similarly, Yoon et al. (2002) used sputtering technique to deposit Pd film on the surface of Nafion® 117. It was found that the Pd films thinner than 300 Å were dense and appeared to be well attached to the membrane, but there were many cracks in the 1000 Å films. The 1000 Å films were very unstable and were easily delaminated from the membrane surface. When the composite membrane is immersed in water, the Nafion® membrane swells very much, but the Pd film can not expand as much as the membrane (Yoon et al., 2002). The proton conductivity was found to decrease with increasing Pd thickness. For the Pd-1000 Å film, 30% reduction in conductivity was observed. Methanol permeability at 25°C decreased with increasing Pd thickness and they varied from 2.90×10^{-6} to 2.23×10^{-6} cm^2 s^{-1} by deposition of Pd film of 1000 Å on the Nafion® 117 membrane. For the Pd-1000 Å-Nafion® 115 membrane, permeability decreased as much as 44% from 2.97×10^{-6} to 1.67×10^{-6} cm^2 s^{-1}. The cell performance at 90°C exhibited a slight decrease with the Pd layered Nafion® 115 membranes. The methanol crossover through an MEA is inversely proportional to current density and thus, its effect on the performance is more prominent at low current densities. It was found that at low current densities cell performance increased in Pd film of 1000 Å on the Nafion® 115, where the Pd film act as a barrier to methanol crossover. However, these results are different of that of Choi et al. who observed significant increase in DMFC performance in Pd-20 nm film on Nafion® 117.

Z.Q. Ma et al. (2003) followed a different approach, where Nafion® membrane was modified by sputtering a thin layer of Pt/Pd-Ag/Pt on its surface. The methanol crossover can be reduced by sputtering Pd-Ag alloys over the polymer electrolyte, furthermore, when hydrogen is absorbed and dissolved in the membrane, the palladium-silver alloy membrane not only reduces the possibility of embrittling due to α → β phase transition at low temperatures (< 150°C) but also leads to a higher permeability for hydrogen than pure palladium membrane. The composite membranes were prepared as follows: on one side of the Nafion® 117 membrane, a 2 nm Pt film was first deposited. This was followed by a Pd-Ag film with three different thicknesses (0.1, 0.2 and 1 μm). On the top of the Pd-Ag, it was coated with another 2 nm Pt film. Before a MEA was manufactured, a 4-5 μm layer of Nafion® polymer was recast over the surface of the sputtered Pt/Pd-Ag/Pt layer with Nafion® solution. The final membrane prepared was in the form of Nafion®117/Pt/Pd-Ag/Pt/Nafion®, containing 0.0086 mg cm^{-2} Pt, 0.90 mg cm^{-2} Pd, and 0.27 mg cm^{-2} Ag. The cell performance and the OCV increased with increasing the sputtering alloy layer thickness and the best performance and the highest OCV were found with the 1 μm Pd-Ag film. Also it was showed that the performance with the 1 μm Pd-Ag film is higher than that of cell with a Nafion® membrane having catalyst loading twice as high. Palladinized Nafion® composite membrane was prepared via ion-exchange and chemical reduction method (Y.J. Kim et al., 2004). Palladium(II) acetylacetonate and tetraammine-palladium(II) chloride hydride were used as palladium precursors. Nafion® 117 samples were immersed in palladium precursor solutions followed by chemical reduction of palladium precursors by sodium borohydride. The use of tetraamminepalladium(II) chloride hydride formed 40-50 nm of palladium particles while palladium(II) acetylacetonate formed 5-10 nm of particles. For all palladinized samples, water uptake was higher than for unfilled Nafion® whereas above a certain amount of incorporated palladium, methanol uptake was lower than for unfilled Nafion®. Incorporating Pd nanoparticles decreases the proton conductivity and methanol permeability, compared to bare Nafion®, simultaneously. However, the conductivity increased and methanol permeability decreased as the amount of incorporated Pd increased, and above a certain amount the rapid increase of conductivity and permeability appeared. The DMFC performance at 40°C was improved by incorporating Pd.

Different approach for the incorporation of Pd in Nafion® membranes was suggested by Tang et al. (2005). Multi-layer self-assembly Nafion® membranes (MLSA Nafion® membranes) were prepared by alternately assembling charged Pd particles and Nafion®

ionomer onto Nafion® 112 membranes. The Pd particles, size of about 1.8 nm in average, are charged by PDDA (polydiallyldimethylammonium chloride) ionomers. The Pd loading of the first-layer MLSA Nafion® membranes was 0.63 μg cm^{-2}, and the surface coverage of the Pd nanoparticles on the Nafion® membrane was estimated as 22%. After 5-double-layer Pd particles/Nafion® ionomers assembling, the Pd loading reached to 2.86 μg cm^{-2}. The methanol crossover current of the original Nafion® membranes and 1-double-layer, 2-double-layer, 3-double-layer, 4-double-layer, 5-double-layer MLSA Nafion® membrane were 0.0495, 3.87 × 10^{-3}, 1.38 × 10^{-3}, 7.32 × 10^{-4}, 5.16 × 10^{-4} and 4.25 × 10^{-3} A cm^{-2}, respectively, corresponding conductivities of 0.112, 0.110, 0.105, 0.094, 0.087 and 0.081 S cm^{-2}. No DMFC data were provided, however, it was suggested that the 3-double-layer self-assembly membrane is the best suited for DMFC application, since it has a methanol crossover decreased to 0.86%, and a conductivity remaining at 83.9% comparing to original Nafion® membrane. Electroless plating was also used to deposit Pd layer on Nafion® membranes (Hejze et al., 2005; Sun et al., 2005). Palladium layer can reduce methanol crossover when coated on the surface of Nafion®, i.e. limiting current from methanol permeation through a membrane electrode assembly was reduce from 64 to 57 mA cm^{-2} for 1M methanol, and from 267 to 170 mA cm^{-2} for 5M methanol, for Nafion® 115 and Pd/Nafion® 115 membranes respectively (Sun et al., 2005). Also it was demonstrated that the DMFC performance increase with the incorporation of Pd in Nafion® 115. When 1M methanol was used, the power density increased from 36 to 45 mW cm^{-2}, for Nafion® 115 and Pd/Nafion® 115, respectively. When 5M methanol was used, the maximum power density on Pd/Nafion® 115 was 72 mW cm^{-2}, while the performance of MEA with pure Nafion® 115 membranes was only 32 mW cm^{-2}.

6.7 Organic/montmorillonite nanocomposite membranes

Montmorillonite (MMT) is a type of layered silicate composed of silica tetrahedral and alumina octahedral sheets (J. Chang et al., 2003) and its intercalation into Nafion® membrane can decrease successfully the methanol permeability and improve mechanical property (Jung et al., 2003; Song et al., 2004). Research by J. Chang et al. (2003) has showed that layered silicates incorporated into SPEEK membranes helped to reduce swelling significantly in hot water and decrease the methanol crossover without a serious reduction of the proton conductivity. Gaowen and Zhentao (2005) prepared organically modified MMT (OMMT) through ion exchange reaction between alkylammonium cations and metal cations. The nanocomposite membranes (SPEEK/OMMT) were prepared using the solution intercalation technique. The water uptake of SPEEK membrane increased rapidly above 50°C, while the SPEEK/OMMT composite membranes posses of rather constant water uptake up to 80°C. This indicates that MMT layers incorporated into SPEEK matrix prevent extreme swelling of the composite membranes due to the cohesion of the functional groups between SPEEK matrix and MMT layers. The proton conductivity of the membrane was measured at temperature ranging from 22 to 110°C. It was found that the conductivity of SPEEK/OMMT composite is lower than that of the pristine SPEEK and decreases sequentially as the content of OMMT increases, which is due to prolonging the transfer route of proton. The conductivity of SPEEK/OMMT (5 wt.%) approaches the value of Nafion® 115 at 90°C and reaches 1.2 × 10^{-2} S cm^{-1}. The activation energies of SPEEK/OMMT are higher than that of Nafion® 115, where the value of 32.08 kJ/mol and 10.8 kJ/mol, respectively, were found. Methanol permeability was found to be in the following order: Nafion® 115> SPEEK > SPEEK/OMMT. Therefore, incorporating nano-sized dispersion of OMMT prevents methanol from migrating through the membrane. Song *et al.* (2004) prepared recast Nafion®/MMT nanocomposite membranes at a loading of 3 wt.% clay. It was found that the strength increased more than 35% and the tensile elongation almost doubled. The thermal decomposition behavior of Nafion®/MMT nanocomposite was similar to that of pristine Nafion®, but the major decomposition temperature of polymer main chain shifted to much higher temperature region. The proton conductivity of pristine recast Nafion® with thickness of ca. 100 μm and dry state reached about 0.08 S/cm at room temperature. For Nafion®/MMT nanocomposite membranes, the room temperature conductivity was almost similar to that of neat Nafion® below MMT loading of 2 wt. % and then decreased gradually with the increase in filler content. The methanol permeability of pristine recast Nafion® was 2.3×10^{-6} cm³ cm/ cm² s, while for Nafion®/MMT composite membranes with a thickness of 50 μm significantly decreases to 1.6×10^{-7} cm³ cm/ cm² s by only 1 wt.% organo clay loading, which amounted to more than 90% reduction. Jung *et al.* (2003) prepared dodecylamine-exchanged montmorllonite (m-MMT) by a cation exchange reaction. The thermal resistance of Nafion®/MMT composite was found to be lower than that of the pristine Nafion®. Also the thermal resistance of 5 and 7 wt.% MMT was lower than that of 3 wt.% MMT. On the other hand, Nafion®/MMT displayed higher thermal resistance than that of pristine Nafion®. The thermal resistance of the Nafion®/m-MMT nanocomposite was also increased slightly with increasing the contents of m-MMT in the composite membrane. The methanol permeability of pristine Nafion® was found to be 0.13 mol/l at 1 h. By adding MMT and m-MMT, the methanol permeability decreased to 0.045 and 0.042 mol/l at 1 h, respectively. The proton conductivity of Nafion®/MMT was found to be 8.9×10^{-2}, 7×10^{-2}, 7.2×10^{-2} and 6.7×10^{-2} S/cm at 110°C for a content of MMT of 0, 3, 5 and 7 wt.%, respectively. However, the conductivity of Nafion®/m-MMT was found to be 8.9×10^{-2}, 7.72×10^{-2}, 7.57×10^{-2} and 7.4×10^{-2} S/cm at 110°C for a content of m-MMT of 0, 3, 5 and 7 wt.%, respectively. In general the proton conductivity of the composite membranes decreased slightly with increasing the contents of MMT and m-MMT and lower than pristine Nafion®. Pristine Nafion® performances were 385, 410 and 138.1 mA/cm² (at a potential of 0.4 V) at 90, 110 and 125°C, respectively. For Nafion®/3 wt.% MMT performances were 370, 452.6 and 282.86 mA/cm² (at a potential of 0.4 V) at 90, 110 and 125°C, respectively. Finally, for Nafion®/3 wt.% m-MMT performances were 367.1, 440 and 290 mA/cm² (at a potential of 0.4 V) at 90, 110 and 125°C, respectively. Gosalawit *et al.* (2008) prepared sulfonated montmorillonite (SMMT) with SPEEK. It was found that the inorganic aggregation in SPEEK increased with SMMT loading. The stability in water and in methanol aqueous solution as well as the mechanical stability were enhanced with SMMT loading. Whereas thermal stability improvement did not exist significantly. The methanol permeability was reduced when the SMMT loading increased. The proton conductivity was improved with the incorporation of SMMT. SMMT/SPEEK nanocomposite membranes showed significant cell performance for DMFC as compared to pristine SPEEK and Nafion® 117 membranes.

6.8 Organic/zeolites nanocomposite membranes

Mordenite crystals dispersed in poly(acrylic acid) was prepared by Rao *et al.* (1994) and Libby *et al.* (2001), while mordenite dispersed in Nafion® was prepared by Arimura *et al.* (1999). These membranes displayed proton conductivity about two orders of magnitude lower than that of pristine Nafion®. A study of using ZSM5 as a filler in Nafion® was study by Byun et al. (2006). These composite membranes show higher water uptake than Nafion® 115, while methanol permeability has decrease with increasing zeolite contents. The selectivity of nanocomposite membranes was higher than that of Nafion® 115. Zeolite beta was incorporated in Nafion® (Holmberg et al., 2008) and Chitosan (Y. Wang et al., 2008) membranes. Zeolite beta/ Nafion® nanocomposite membranes with loading of 2.5 and 5 wt.% posses proton conductivity/methanol permeability (selectivity) ratios as much as 93% higher than commercial Nafion® 117 at 21°C, and 63% higher at 80°C. These composite membranes outperform Nafion® 117 in DMFC (Holmberg et al., 2008). The incorporation of zeolite beta in Chitosan reduces the methanol permeability. Furthermore, zeolite beta was sulfonated, therefore the methanol permeability was further reduced as a result of the enhanced interfacial interaction between zeolite beta and Chitosan matrix. The cell performance of the composite membrane were comparable to Nafion® 117 at low methanol concentration (2M) and much better at high methanol concentration (12M) (Y. Wang et al., 2008). Other zeolite based nanocomposite membranes investigated are MCM-41 (Bello et al., 2008; Gomes et al., 2008; Karthikeyan et al., 2005; Marschall et al., 2007), zeolite Y (Ahmad et al., 2006), zeolite BEA (Holmberg et al., 2005) and Chabazite and clinoptilolite (Tricoli & Nannetti, 2003). These membranes can maintained the proton conductivity at temperature above 100°C, also methanol permeability is reduced by incorporating these zeolites.

7. Conclusion

High temperature operating fuel cell enhanced the performance, especially for methanol, DME and ethanol fuel cells. Incorporating nano inorganic materials in the organic matrix has several advantages, namely increase membrane thermal and mechanical stabilities, increase the working temperature and water retention. Some inorganic fillers are proton conductor which can increase the conductivity of the composite membrane or at least keep the same conductivity of the pristine organic membrane. Fuel crossover also can be reduced by incorporating these inorganic nano materials.

From all the investigated inorganic fillers, the nano size played a major role to enhance the compatibility and the interaction of the inorganic fillers with the polymeric matrix. Also the optimal inorganic loading was found to be around 5 wt.%, with the majority around 3 wt.%.

However, for commercialization of these nanocomposite membranes, more R&D needed to be done, which include:

- Systematic study of these nanocomposite, especially inorganic loading;
- The interaction between the nano-inorganic materials with the organic matrix need to be understood;
- High temperature fuel cell performance needed to be done with very little or no humidification and no pressure;
- Long term fuel cell performance;
- Long term membrane stability (thermal and mechanical) and also long term membrane leaching; and
- A comprehensive comparative study between all the investigated inorganic materials and the interaction with different organic materials.

8. References

Adamson, K.-A. & Pearson, P. (2000). *J. Power Sources*, 86, pp. 548-555

Adjemian, K.T.; Lee, S.J.; Srinivasan, S.; Benziger, J. & Bocarsly, A.B. (2002a). *J. Electrochem. Soc.*, 149, pp. A256-A261

Adjemian, K.T.; Srinivasan, S.; Benziger, J. & Bocarsly, A.B. (2002b). *J. Power Sources*, 109, pp. 356-364

Agrell, J.; Hsselbo, K.; Jansson, K.; Jaras, S.G. & Boutonnet, M. (2001). *Appl. Catal. A*, 211, pp. 239-250

Aharoni, S.M. & Litt, M.H. (1974). *J. Polym. Sci., Polym. Chem. Ed.*, 12, pp. 639-650

Ahmad, M.I.; Zaidi, S.M.J. & Rahman, S.U. (2006). *Desalination*, 193, pp. 387-397

Alberti, G.; Casiola, M.; Costantino, U.; Levi, G. & Ricciard, G. (1978). *J. Inorg. Nucl. Chem.*, 40, pp. 533-537

Alberti, G.; Casciola, M.; Massinelli, L. & Bauer, B. (2001). *J. Membr. Sci.*, 185, pp. 73-81

Alberti, G.; Casciola, M.; D`Alessandro, E. & Pica, M. (2004). *J. Mater. Chem.*, 14, pp. 1910-1914

Alberti, G.; Casciola, M.; Donnadio, A.; Piaggio, P.; Pica, M. & Sisani, M. (2005a). *Solid State Ionics*, 176, pp. 2893-2898

Alberti, G.; Pica, M. & Tarpanelli, T. (2005b). PCT Patent WO 2005/105667 A1

Alberti, G.; Casiola, M.; Capitani, D.; Donnadio, A.; Narducci, R.; Pica, M. & Sganappa, M. (2007). *Electrochim. Acta*, 52, pp. 8125-8132

Allcock, H.R.; Hofmann, M.A.; Ambler, C.M. & Morford, R.V. (2002a). *Macromolecules*, 35, pp. 3484-3489

Allcock, H.R.; Hofmann, M.A.; Ambler, C.M.; Lvov, S.N.; Zhou, X.Y.; Chalkova, E. & Weston, J. (2002b). *J. Membr. Sci.*, 201, pp. 47-54

Amphlett, J.C.; Evans, M.J.; Mann, R.F. & Weir, R.D. (1985). *Can. J. Chem. Eng.*, 63, pp. 605-611

Amphlett, J.C.; Peppley, B.A.; Halliop, E. & Sadiq, A. (2001). *J. Power Sources*, 96, pp. 204-213

Anantaraman, A.V. & Gardner, C.L. (1996). *J. Electroanal. Chem.*, 414, pp. 115-120

Antonucci, P.L.; Aricò, A.S.; Cretì, P.; Ramunni, E. & Antonucci, V. (1999). *Solid State Ionics*, 125, pp. 431-437

Appleby, A.J. & Yeager, E.B. (1986a). In: *Assessment of Research Needs for Advanced Fuel Cells*, Penner, S.S. (Ed.), Chapter 4, Pergamon, New York

Appleby, A.J. & Yeager, E.B. (1986b). *Energy*, 11, pp. 137-151

Argyropoulos, A.; Scott, K. & Taama, W.M. (1999a). *J. Appl. Electrochem.*, 29, pp. 661-669

Argyropoulos, P.; Scott, K. & Taama, W.M. (1999b). *Electrochim. Acta*, 44, pp. 3575-3584

Aricò, A.S.; Cretì, P.; Antonucci, P.L. & Antonucci, V. (1998). *Electrochem. Solid State Lett.*, 1, pp. 66-68

Aricò, A.S.; Baglio, V.; Di Blasi, A. & Antonucci, V. (2003a). *Electrochem. Communications*, 5, pp. 882-866

Aricò, A.S.; Baglio, V.; Di Blasi, A.; Cretì, P.; Antonucci, P.L. & Antonucci, V. (2003b). *Solid State Ionics*, 161, pp. 251-265

Aricò, A.S.; Baglio, V.; Di Blasi, A.; Modica, E.; Antonucci, P.L. & Antonucci, V. (2004). *J. Power Sources*, 128, pp. 113-118

Arimura, T.; Ostrovskii, D.; Okada, T. & Xie, G. (1999). *Solid State Ionics*, 118, pp. 1-10

Baglio, V.; Arico, A.S.; Di Blasi, A.; Antonucci, V.; Antonucci, P.L.; Licoccia, S.; Traversa, E. & Serraino Fiory, F. (2005). *Electrochim. Acta*, 50, pp. 1241-1246

Bailly, C.; Williams, D.J.; Karasz, F.E. & MacKnight, W.J. (1987). *Polymer*, 28, pp. 1009-1016

Baradie, B.; Dodelet, J.P. & Guay, D. (2000). *J. Electroanal. Chem.*, 489, pp. 101-105

Baradie, B.; Poinsignon, C.; Sanchez, J.Y.; Piffard, Y.; Vitter, G.; Bestaoui, N.; Foscallo, D.; Denoyelle, A.; Delabouglise, D. & Vaujany, M. (1998). *J. Power Sources*, 74, pp. 8-16

Bauer, B.; Menzel, Th. & Kehl, P. (1994). *EP 0 645 175 A1*

Bauer, B.; Rafler, G. & Ulrich, H.-H. (1995). *Ger. Patent DE 195 38 025*

Bauer, B.; Jones, D.J.; Rozière, J.; Tchicaya, L.; Alberti, G.; Casciola, M.; Massinelli, L.; Peraio, A.; Besse, S. & Ramunni, E. (2000). *J. New. Mater. Electrochem. Systems*, 3, pp. 93-98

Bauer, F. & Willert-Porada, M. (2004). *J. Membr. Sci.*, 233, pp. 141-149

Bauer, F. & Willert-Porada, M. (2005). *J. Power Sources*, 145, pp. 101-107

Bauer, F. & Willert-Porada, M. (2006a). *Fuel Cells*, 6, pp. 261-269

Bauer, F. & Willert-Porada, M. (2006b). *Solid State Ionics*, 177, pp. 2391-2396

Bello, M.; Javaid Zaidi, S.M. & Rahman, S.U. (2008). *J. Membr. Sci.*, 322, pp. 218-224

Belyakov, V.N. & Linkov, V.M. (1999). *US Patent 5,932,361*

Bernardi, D.M. & Verbrugge, M.W. (1991). *AIChE Journal*, 37, pp. 1151-1163

Bi, C.; Zhang, H.; Zhang, Y.; Zhu, X.; Ma, Y.; Dai, H. & Xiao, S. (2008). *J. Power Sources*, 184, pp. 197-203

Boffito, G. (1999). *US Patent 5,976,723*

Bonnet, B.; Jones, D.J.; Rozière, J.; Tchicaya, L.; Alberti, G.; Casciola, M.; Massinelli, L.; Bauer, D.; Peraio, A. & Ramunni, E. (2000). *J. New Mater. Electrochem. Systems*, 3, pp. 87-92

Brack, H.P.; Wyler, M.; Peter, G. & Scherer, G.G. (2003). *J. Membr. Sci.*, 214, pp. 1-19

Breen, J.P. & Ross, J.R.H. (1999). *Catalysis Today*, 51, pp. 521-533

Büchi, F.N.; Gupta, B.; Haas, O. & Scherer, G.G. (1995a). *Electrochim. Acta*, 40, pp. 345-353

Büchi, F.N.; Gupta, B.; Haas, O. & Scherer, G.G. (1995b). *J. Electrochem. Soc.*, 142, pp. 3044-3048

Byun, S.C.; Jeung, Y.J.; Park, J.W.; Kim, S.D.; Ha, H.Y. & Kim, W.J. (2006). *Solid State Ionics*, 177, pp. 3233-3243

Carrette, L.; Friedrich, K.A. & Stimming, U. (2001). *Fuel Cells*, 1, pp. 5-39

Casiola, M.; Alberti, G.; Ciarletta, A.; Cruccolini, A.; Piaggio, P. & Pica, M. (2005). *Solid State Ionics*, 176, pp. 2985-2989

Casiola, M.; Alberti, G.; Sganappa, M. & Narducci, R. (2006). *J. Power Sources*, 162, pp. 141-145

Casiola, M.; Capitani, D.; Comite, A.; Donnadio, A.; Frittella, V.; Pica, M.; Sganappa, M. & Varzi, A. (2008). *Fuel Cells*, 8, pp. 217-224

Chang, H.Y. & Lin, C.W. (2003). *J. Membr. Sci.*, 218, pp. 295-306

Chang, J.; Park, J.H.; Park, G.; Kim, C. & Park, O. (2003). *J. Power Sources*, 124, pp. 18-25

Choi, W.C.; Kim, J.D. & Woo, S.I. (2001). *J. Power Sources*, 96, pp. 411-414

Chu, D. & Gilman, S. (1994). *J. Electrochem. Soc.*, 141, pp. 1770-1773

Chu, D. & Jiang, R. (1999). *J. Power Sources*, 80, pp. 226-234

Colbow, K.M.; Zhang, J. & Wilkinson, D.P. (2000). *J. Electrochem. Soc.*, 147, pp. 4058-4060

Colicchio, I.; Wen, F.; Keul, H.; Simon, U. & Moeller, M. (2009). *J. Membr. Sci.*, 326, pp. 45-57

Cornet, N.; Diat, O.; Gebel, G.; Jousse, F.; Marsaq, D.; Mercier, R. & Peneri, M. (2000). *J. New Mater. Electrochem. Systems*, 3, pp. 33-42

Costamagna, P.; Yang, C.; Bocarsly, A.B. & Srinivasan, S. (2002). *Electrochim. Acta*, 47, pp. 1023-1033

Cubiero, M.L. & Fierro, J.L.G. (1998). *J. Catalysis*, 179, pp. 150-162

Cui, W.; Kerres, J. & Eigenberger, G. (1998). *Sep. Purif. Technol.*, 14, pp. 145-154

D`Alelio, G. (1944). *US Patent 2,340,111*

Deng, Q.; Moore, R.B. & Mauritz, K.A. (1998). *J. Appl. Polym. Sci.*, 68, pp. 747-763

Di Vona, L.M.; Ahmed, Z.; Bellitto, S.; Lenci, A.; Traversa, E. & Licoccia, S. (2007). *J. Membr. Sci.*, 296, pp. 156-161

Dimitrova, P.; Friedrich, K.A.; Stimming, U. & Vogt, B. (2002a). *Solid State Ionics*, 150, pp. 115-122

Dimitrova, P.; Friedrich, K.A.; Vogt, B. & Stimming, U. (2002b). *J. Electroanal. Chem.*, 532, pp. 75-83

Dohle, H.; Divisek, J.; Mergel, J.; Oetjen, H.F.; Zingler, C. & Stolten, D. (2002). *J. Power Sources*, 105, pp. 274-282

Duesterwald, H.G.; Höhlein, B.; Kraut, H.; Meusinger, J.; Peters, R. & Stimming, U. (1997). *Chem. Eng. Technol.*, 20, pp. 617-623

Edwards, N.; Ellis, S.R.; Frost, J.C.; Golunski, S.E.; van Keulen, A.N.J.; Lindewald, N.G. & Reinkingh, J.G. (1998). *J. Power Sources*, 71, pp. 123-128

Emonts, B.; Hansen, J.B.; Jørgensen, S.L.; Höhlein, B. & Peters, R. (1998). *J. Power Sources*, 71, pp. 288-293

Fan, J.; Guo, X.; Harada, S.; Watari, T.; Tanaka, K.; Kita, H. & Okamoto, K. (2002). *Macromolecules*, 35, pp. 9022-9028

Faungnawakij, K. & Viriya-empikul, N. (2010). *Appl. Cat. A*, 382, pp. 21-27

Faungnawakij, K.; Shimoda, N.; Viriya-empikul, N.; Kikuchi, R. & Eguchi, K. (2010). *Appl. Cat. B*, 97, pp. 21-27

Faure, S.; Mercier, R.; Aldebert, P.; Pineri, M. & Sillion, B. (1996). *Fr. Patent* 96 05707

Faure, S.; Cornet, N.; Gebel, G.; Mercier, R.; Pineri, M. & Sillion, B. (1997). In: *Proceeding of the Second International Symposium on New Materials for Fuel Cell and Modern Battery Systems*, Savadogo, O. & Roberge, P.R. (Eds.), pp. 818, Montréal, Canada, July 6-10, 1997

Ferrell, J.R.; Kuo, M.C. & Herring, A.M. (2010). *J. Power Sources*, 195, pp. 39-45

Finsterwalder, F. & Hambitzer, G. (2001). *J. Membr. Sci.*, 185, pp. 105-124

Fontanella, J.J.; Wintersgill, M.C.; Chen, R.S.; Wu, Y. & Greenbaum, S.G. (1995). *Electrochim. Acta*, 40, pp. 2321-2326

Fujiwara, N.; Friedrich, K.A. & Stimming, U. (1999). *J. Electroanal. Chem.*, 472, pp. 120-125

Gaowen, Z. & Zhentao, Z. (2005). *J. Membr. Sci.*, 261, pp. 107-113

Gavach, C.; Pamboutzoglou, G.; Nedyalkov, M. & Pourcelly, G. (1989). *J. Membr. Sci.*, 45, pp. 37-53

Gebel, G.; Aldebert, P. & Pineri, M. (1993). *Polymer*, 34, pp. 333-339

Genies, C.; Mercier, R.; Sillion, B.; Cornet, N.; Gebel, G. & Pineri, M. (2001). *Polymer*, 42, pp. 359-373

Giordano, N.; Staiti, P.; Hocevar, S. & Arico, A.S. (1996). *Electrochim. Acta*, 41, pp. 397-403

Glazebrook, R.W. (1982). *J. Power Sources*, 7, pp. 215-256

Glipa, X.; El Haddad, M.; Jones, D.J. & Rozière, J. (1997). *Solid State Ionics*, 97, pp. 323-331

Gloaguen, F.; Convert, P.; Gamburzev, S.; Velev, O.A. & Srinivasan, S. (1998). *Electrochim. Acta*, 43, pp. 3767-3772

Gode, P.; Ihonen, J.; Strandroth, A.; Ericson, H.; Lindbergh, G.; Paronen, M.; Sundholm, F.; Sundholm, G. & Walsby, N. (2003). *Fuel Cells*, 3, pp. 21-27

Gomes, D.; Marschall, R.; Nunes, S.P. & Wark, M. (2008). *J. Membr. Sci.*, 322, pp. 406-415

Gong, X.; Bandis, A.; Tao, A.; Meresi, G.; Wang, Y.; Inglefield, P.T.; Jones, A.A. & Wen, W.-Y. (2001). *Polymer*, 42, pp. 6485-6492

Goodstein, E.S. (1999). *Economics and Environment*, Upper Saddle River, NJ, Prentice Hall

Gosalawit, R.; Chirachanchai, S.; Shishatskiy, S. & Nunes, S.P. (2008). *J. Membr. Sci.*, 323, pp. 337-346

Gottesfeld, S. & Zawodzinski, T.A. (1997). In: *Advances in Electrochemical Science and Enginering*, Alkire, R.C.; Gerischer, H.; Kolb, D.M. & Tobias, C.W. (Eds.), Vol. 5, Chapter 4, Wiley-VCH, Weinheim

Grot, W. (1978). *Chem. Ing. Tech.*, 50, pp. 299-301; (1975). *Chem. Ing. Tech.*, 47, pp. 617; (1972). *Chem. Ing. Tech.*, 44, pp. 167-169

Grot, W.G. & Rajendran, G. (1999). *US Patent* 5,919,583

Guo, Q.; Pintauro, P.N.; Tang, H. & O'Connor, S. (1999). *J. Membr. Sci.*, 154, pp. 175-181

Gupta, B.; Büchi, F.N.; Scherer, G.G. & Chapiró, A. (1993). *Solid State Ionics*, 61, pp. 213-218

Gupta, B.; Büchi, F.N. & Scherer, G.G. (1994). *J. Polym. Sci. A: Polym. Chem.*, 32, pp. 1931-1938

Gupta, B. & Scherer, G.G. (1994). *Chimia*, 48, pp. 127-137

Hacker, V. & Kordesch, K. (2003). In: *Handbook of Fuel Cells*, Vielstich, W.; Lamm, A. & Gasteiger, H.A. (Eds.), Vol 3, Part 1, Chapter 10, pp. 121-127, Wiley, Chichester, England

Hara, S. & Miyayama, M. (2004). *Solid State Ionics*, 168, pp. 111-116

Hatanaka, T.; Hasegawa, N.; Kamiya, A.; Kawasumi, M.; Morimoto, Y. & Kawahara, K. (2002). *Fuel*, 81, pp. 2173-2176

Hejze, T.; Gollas, B.R.; Sauerbrey, R.K.; Schmied, M.; Hofer, F. & Besenhard, J.O. (2005). *J. Power Sources*, 140, pp. 21-27

Helen, M.; Viswanathan, B. & Srinivasa Murthy, S. (2006). *J. Power Sources*, 163, pp. 433-439

Helen, M.; Viswanathan, B. & Srinivasa Murthy, S. (2007). *J. Membr. Sci.*, 292, pp. 98-105

Hill, M.L.; Kim, Y.S.; Einsla, B.R. & McGrath, J.E. (2006). *J. Membr. Sci.*, 283, pp. 102-108

Hoel, D. & Grunwald, E. (1977). *J. Phys. Chem.*, 81, pp. 2135-2136

Hogarth, M.; Christensen, P.; Hamnett, A. & Shukla, A. (1997). *J. Power Sources*, 69, pp. 113-124

Hogarth, M.P. & Ralph, T.R. (2002). *Platinum Metals Rev.*, 46, pp. 146-164

Höhlein, B.; Boe, M.; Bøgild-Hansen, J.; Bröckerhoff, P.; Colsman, G.; Emonts, B.; Menzer, R. & Riedel, E. (1996). *J. Power Sources*, 61, pp. 143-147

Holmberg, B.A.; Hwang, S.J.; Davis, M.E. & Yan, Y. (2005). *Microp. Mesop. Mat.*, 80, pp. 347-356

Holmberg, B.A.; Wang, X. & Yan, Y. (2008). *J. Membr. Sci.*, 320, pp. 86-92

Honma, I.; Nishikawa, O.; Sugimoto, T.; Nomura, S. & Nakajima, H. (2002). *Fuel Cells*, 2, pp. 52-58

Hou, H.; Sun, G.; Wu, Z.; Jin, W. & Xin, Q. (2008). *Int. J. Hydrogen Energy*, 33, pp. 3402-3409

Hubner, G. & Roduner, E. (1999). *J. Mater. Chem.*, 9, pp. 409-418

Ioannides, T. & Neophytides, S. (2000). *J. Power Sources*, 91, pp. 150-156

Jiang, C.J.; Trimm, D.L.; Wainwright, M.S. & Cant, N.W. (1993). *Appl. Cat. A*, 93, pp. 245-255

Jiang, R.; Kunz, H.R. & Fenton, J.M. (2006a). *Electrochim. Acta*, 51, pp. 5596-5605

Jiang, R.C.; Kunz, H.R. & Fenton, J.M. (2006b). *J. Membr. Sci.*, 272, pp. 116-124

Jian-hua, T.; Peng-fei, G.; Zhi-yuan, Z.; Wen-hui, L. & Zhong-qiang, S. (2008). *Int. J. Hydrogen Energy*, 33, pp. 5686-5690

Johnson, B.C.; Yilgor, I.; Tran, C.; Iqbal, M.; Wightman, J.P.; Llyod, D.R. & McGrath, J.E. (1984). *J. Polym. Sci.*, 22, pp. 721-737

Jones, D.J. & Rozière, J. (2001). *J. Membr. Sci.*, 185, pp. 41-58

Jones, D.J. & Rozière, J. (2001). In: *Handbook of Fuel Cells*, Vielstich, W.; Lamm, A. & Gasteiger, H.A. (Eds.), Vol 3, Part 1, Chapter 35, pp. 447-455, Wiley, Chichester, England

Jörissen, L.; Gogel, V.; Kerres, J. & Garche, J. (2002). *J. Power Sources*, 105, pp. 267-273

Jung, D.H.; Cho, S.Y.; Peck, D.H.; Shin, D.R. & Kim, J.S. (2002). *J. Power Sources*, 106, pp. 173-177

Jung, D.H.; Cho, S.Y.; Peck, D.H.; Shin, D.R. & Kim, J.S. (2003). *J. Power Sources*, 118, pp. 205-211

Karthikeyan, C.S.; Nunes, S.P.; Prado, L.A.S.A.; Ponce, M.L.; Silva, H.; Ruffmann, B & Schulte, K. (2005). *J. Membr. Sci.*, 254, pp. 139-146

Kerangueven, G.; Coutanceau, C.; Sibert, E.; Hahn, F.; Leger, J.M. & Lamy, C. (2006). *J. Appl. Electrochem.*, 36, pp. 441-448

Kerres, J.; Cui, W. & Reichle, S. (1996). *J. Polym. Sci.: Part A: Polym. Chem.*, 34, pp. 2421-2438

Kerres, J.; Cui, W. & Schnurnberger, W. (1997). *D. Patent* 19622337.7; (1997). *Fr. Patent* F. 9706706; (1997). *US Patent* 08/868943

Kerres, J. ; Cui, W.; Disson, R. & Neubrand, W. (1998a). *J. Membr. Sci.*, 139, pp. 211-225

Kerres, J.; Cui, W. & Junginger, M. (1998b). *J. Membr. Sci.*, 139, pp. 227-241

Kerres, J.; Zhang, W. & Cui, W. (1998c). *J. Polym. Sci.: Part A: Polym. Chem.*, 36, pp. 1441-1448

Kerres, J.; Ullrich, A.; Meier, F. & Häring, T. (1999). *Solid State Ionics*, 125, pp. 243-249

Kerres, J.; Ullrich, A.; Häring, T.; Baldauf, M.; Gebhardt, U. & Preidel, W. (2000). *J. New Mater. Electrochem. Systems*, 3, pp. 229-239

Kerres, J.A. (2001). *J. Membr. Sci.*, 185, pp. 3-27

Kim, D.S.; Park, H.B.; Rhim, J.W. & Lee, Y.M. (2004). *J. Membr. Sci.*, 240, pp. 37-48

Kim, H.J.; Shul, Y.G. & Han, H. (2006). *J. Power Sources*, 158, pp. 137-142

Kim, Y.J.; Choi, W.C.; Woo, S.I. & Hong, W.H. (2004). *Electrochim. Acta*, 49, pp. 3227-3234

Kim, Y.S.; Dong, L.; Hickner, M.A.; Pivovar, B.S. & McGrath, J.E. (2003). *Polymer*, 44, pp. 5729-5736

Kim, Y.T.; Song, M.K.; Kim, K.H.; Park, S.B.; Min, S.K. & Rhee, H.W. (2004). *Electrochim. Acta*, 50, pp. 645-648

Kobayashi, T.; Rikukawa, M.; Sanui, K. & Ogata, N. (1998). *Solid State Ionics*, 106, pp. 219-225

Kordesch, K. & Simader, G. (1996). In: *Fuel cells and their Applications*, VCH, Weinheim

Kordesch, K.V. & Simader, G.R. (1995). *Chem. Rev.*, 95, pp. 191-207

Kreuer, K.D. (1997). *Solid State Ionics*, 97, pp. 1-15

Kreuer, K.D. (2001). *J. Membr. Sci.*, 185, pp. 29-39

Krishnan, P.; Park, J.S.; Yang, T.H.; Lee, W.Y. & Kim, C.S. (2006). *J. Power Sources*, 163, pp. 2-8

Kuan, H.C.; Wu, C.S.; Chen, C.Y.; Yu, Z.Z.; Dasari, A. & Mai, Y.W. (2006). *Electrochem. Solid State Lett.*, 9, pp. A76-A79

Lamy, C.; Belgsir, E.M. & Léger, J.-M. (2001). *J. Appl. Electrochem.*, 31, pp. 799-809

Larminie, J. & Dicks, A. (2000). In: *Fuel Cell Systems Explained*, Wiley, Chichester

Ledesma, C. & Llorca, J. (2009). *Chem. Eng. J.*, 154, pp. 281-286

Ledjeff-Hey, K.; Formanski, V.; Kalk, Th. & Roes, J. (1998). *J. Power Sources*, 71, pp. 199-207

Lee, H.K.; Kim, J.I.; Park, J.H. & Lee, T.H. (2004). *Electrochim. Acta*, 50, pp. 761-768

Lee, W.; Shibasaki, A.; Saito, K.; Sugita, K.; Okuyama, K. & Sugo, T. (1996). *J. Electrochem. Soc.*, 143, pp. 2795-2798

Léger, J.-M. (2001). *J. Appl. Electrochem.*, 31, pp. 767-771

Lehtinen, T.; Sundholm, G.; Holmberg, S.; Sundholm, F.; Björnbom, P. & Bursell, M. (1998). *Electrochim. Acta*, 43, pp. 1881-1890

Li, C.; Sun, G.; Ren, S.; Liu, J.; Wang, Q.; Wu, Z.; Sun, H. & Jin, W. (2006). *J. Membr. Sci.*, 272, pp. 50-57

Liang, Z.X.; Zhao, T.S. & Prabhuram, J. (2006). *J. Membr. Sci.*, 283, pp. 219-224

Libby, B.; Smyrl, W.H. & Cussler, E.L. (2001). *Electrochem. Solid State Lett.*, 4, pp. A197-A199

Lim, C. & Wang, C.Y. (2003). *J. Power Sources*, 113, pp. 145-150

Lin, C.W.; Thangamuthu, R. & Chang, P.H. (2005). *J. Membr. Sci.*, 254, pp. 197-205

Lin, Y.F.; Yen, C.Y.; Ma, C.C.M.; Liao, S.H.; Lee, C.H.; Hsiao, Y.H. & Lin, H.P. (2007). *J. Power Sources*, 171, pp. 388-395

Linkous, C.A.; Anderson, H.R.; Kopitzke, R.W. & Nelson, G.L. (1998). *Int. J. Hydrogen Energy*, 23, pp. 525-529

Liu, Z.; Guo, B.; Huang, J.; Hong, L.; Han, M. & Ming Gan, L. (2006). *J. Power Sources*, 157, pp. 207-211

Livingston, D.I.; Kamath, P.M. & Corley, R.S. (1956). *J. Polym. Sci.*, 20, pp. 485-490

Lufrano, F.; Squadrito, G.; Patti, A. & Passalacqua, E. (2000). *J. Appl. Polym. Sci.*, 77, pp. 1250-1257

Lufrano, F.; Gatto, I.; Staiti, P.; Antonucci, V. & Passalacqua, E. (2001). *Solid State Ionics*, 145, pp. 47-51

Ma, L.; Jiang, C.; Adesina, A.A.; Trimm, D.L. & Wainwright, M.S. (1996). *Chem. Eng. J.*, 62, pp. 103-111

Ma, Y.-L.; Wainright, J.S.; Litt, M.H. & Savinell, R.F. (2004). *J. Electrochem. Soc.*, 151, pp. A8-A16

Ma, Z.Q.; Cheng, P. & Zhao, T.S. (2003). *J. Membr. Sci.*, 215, pp. 327-336

Marschall, R.; Bannat, I.; Caro, J. & Wark, M. (2007). *Microp. Mesop. Mat.*, 99, pp. 190-196

Mauritz, K.A.; Stefanithis, I.D.; Davis, S.V.; Scheetz, R.W.; Pope, R.K.; Wilkes, G.L. & Huang, H.-H. (1995). *J. Appl. Polym. Sci.*, 55, pp. 181-190

Mauritz, K.A. (1998). *Mater. Sci. Eng. C*, 6, pp. 121-133

Mench, M.; Boslet, S.; Thynell, S.; Scott, J. & Wang, C.Y. (2001). In: *Direct Methanol Fuel Cells*, Narayanan, S.; Zawodzinski, T. & Gottesfeld, S. (Eds.), PV 2001-4, pp. 241, *The Electrochemical Society Proceeding Series*, Pennington, NJ

Mench, M.M.; Chang, H.M. & Wang, C.Y. (2004). *J. Electrochem. Soc.*, 151, pp. A144-A150

Mench, M.W. & Wang, C.Y. (2003). *J. Electrochem. Soc.*, 150, pp. A79-A85

Mokrani, T. & Scurrell, M. (2009). *Cat. Rev., Sci. Eng.*, 51, pp. 1-145

Mitov, S.; Vogel, B.; Roduner, E.; Zhang, H.; Zhu, X.; Gogel, V.; Jorissen, L.; Hein, M.; Xing, D.; Schonberger, F. & Kerres, J. (2006). *Fuel Cells*, 6, pp. 413-424

Mizsey, P.; Newson, E.; Truong, T.-B. & Hottinger, P. (2001). *Appl. Cat. A*, 213, pp. 233-237

Mizutani, I.; Liu, Y.; Mitsushima, S.; Ota, K.I. & Kamiya, N. (2006). *J. Power Sources*, 156, pp. 183-189

Narayanan, S.R.; Kindler, A.; Jeffries-Nakamura, B.; Chun, W.; Frank, H.; Smart, M.; Valdez, T.I.; Surampudi, S.; Halpert, G.; Kosek, J. & Cropley, C. (1996). In: *Proceedings of 11th Annual Battery Conference on Applications and Advances*, pp. 113-122, Long Beach, Calif., Jan. 9-12, 1996

Nilsson, M.; Jozsa, P. & Pettersson, L.J. (2007). *Appl. Cat. B*, 76, pp. 42-50

Nilsson, M.; Jansson, K.; Jozsa, P. & Pettersson, L.J. (2009). *Appl. Cat. B*, 86, pp. 18-26

Nishiguchi, T.; Oka, K.; Matsumoto, T.; Kanai, H.; Utani, K. & Imamura, S. (2006). *Appl. Cat. A*, 301, pp. 66-74

Nolte, R.; Ledjeff, K.; Bauer, M. & Mulhaupt, R. (1993). *J. Membr. Sci.*, 83, pp. 211-220

Nordlund, J.; Roessler, A. & Lindbergh, G. (2002). *J. Appl. Electrochem.*, 32, pp. 259-265

Nunes, S.P.; Ruffmann, B.; Rikowski, E.; Vetter, S. & Richau, K. (2002). *J. Membr. Sci.*, 203, pp. 215-225

Olah, G.A.; Goeppert, A. & Surya Prakash, G.K. (2009). *Beyond oil and Gas: The Methanol Economy*, Wiley-VCH, Weinheim

Papapolymerou, G. & Bontozoglou, V. (1997). *J. Mol. Catal. A*, 120, pp. 165-171

Parthasarathy, A.; Martin, C.R. & Srinivasan, S. (1991). *J. Electrochem. Soc.*, 138, pp. 916-920

Paulus, U.A.; Schmidt, T.J.; Gasteiger, H.A. & Behm, R.J. (2001). *J. Electroanal. Chem.*, 495, pp. 134-145

Powers, E.D. & Serad, G.A. (1986). In: *High Performance Polymers: Their Origin and Development*, Seymour, R.B. & Kirschenbaum, G.S. (Eds.), pp. 355, Elsevier, Amsterdam

Pu, C.; Huang, W.; Ley, K.L. & Smotkin, E.S. (1995). *J. Electrochem. Soc.*, 142, pp. L119-L120

Qi, Z. & Kaufman, A. (2002). *J. Power Sources*, 110, pp. 177-185

Raadschelders, J.W. & Jansen, T. (2001). *J. Power Sources*, 96, pp. 160-166

Ralph, T.R. & Hogarth, M.P. (2002). *Platinum Metals Rev.*, 46, pp. 117-135

Ralph, T.R. (1997). *Platinum Metals Rev.*, 41, pp. 102-113

Ramani, V.; Kunz, H.R. & Fenton, J.M. (2004). *J. Membr. Sci.*, 232, pp. 31-44

Ramani, V.; Kunz, H.R. & Fenton, J.M. (2005a). *Electrochim. Acta*, 50, pp. 1181-1187

Ramani, V.; Kunz, H.R. & Fenton, J.M. (2005b). *J. Membr. Sci.*, 266, pp. 110-114

Rao, N.; Anderson, T.P. & Ge, P. (1994). *Solid State Ionics*, 72, pp. 334-337

Ren, S.; Sun, G.; Li, C.; Song, S.; Xin, Q. & Yang, X. (2006). *J. Power Sources*, 157, pp. 724-726

Ren, X. & Gottesfeld, S. (2001). *J. Electrochem. Soc.*, 148, pp. A87-A93

Rosenblatt, E. & Cohn, J. (1952). *US Patent* 2,601,221

Ruffmann, B.; Silva, H.; Schulte, B. & Nunes, S.P. (2003). *Solid State Ionics*, 162-163, pp. 269 -275

Sacca, A.; Carbone, A.; Passalacqua, E.; D'Epifanio, A.; Licoccia, S.; Traversa, E.; Sala, E.; Traini, F. & Ornelas, R. (2005). *J. Power Sources*, 152, pp. 16-21

Sacca, A.; Gatto, I.; Carbone, A.; Pedicini, R. & Passalacqua, E. (2006). *J. Power Sources*, 163, pp. 47-51

Sambandam, S. & Ramani, V. (2007). *J. Power Sources*, 170, pp. 259-267

Samms, S.R.; Wasmus, S. & Savinell, R.F. (1996). *J. Electrochem. Soc.*, 143, pp. 1225-1232

Santarelli, M.G.L.; Calì, M. & Bertonasco, A. (2003). *Energy Conversion and Management*, 44, pp. 2353-2370

Savadogo, O. (1998). *J. New Mater. Electrochem. Systems*, 1, pp. 47-66

Savadogo, O. & Xing, B. (2000). *J. New Mater. Electrochem. Systems*, 3, pp. 343-347

Savadogo, O. (2004). *J. Power Sources*, 127, pp. 135-161

Schatter, M.J. (1983). In: *Fuel Cells*, Young, G.J. (Ed.), Vol. 2, pp. 290, Reinhold, New York

Scherer, G.G. (1990). *Ber. Bunsenges. Phys. Chem.*, 94, pp. 1008-1014

Schneller, A.; Ritter, H.; Ledjeff, K.; Nolte, R. & Thorwirth, R. (1993). *EP 0574791 A2*

Scott, K.; Taama, W.M. & Argyropoulos, P. (1998). *J. Appl. Electrochem.*, 28, pp. 1389-1397

Serov, A. & Kwak, C. (2009). *Appl. Cat. B*, 91, pp. 1-10

Shanmugam, S.; Viswanathan, B. & Varadarajan, T.K. (2006). *J. Membr. Sci.*, 275, pp. 105-109

Shao, Z.-G.; Joghee, P. & Hsing, I.-M. (2004). *J. Membr. Sci.*, 229, pp. 43-51

Shikada, T.; Asanuma, M. & Ikariya, T. (1991). *US Patent* 5,055,282

Shin, S.-J.; Lee, J.-K.; Ha, H.-Y.; Hong, S.-A.; Chun, H.-S. & Oh, I.-H. (2002). *J. Power Sources*, 106, pp. 146-152

Silva, V.S.; Ruffmann, B.; Silva, H.; Gallego, Y.A.; Mendes, A.; Madeira, L.M. & Nunes, S.P. (2005a). *J. Power Sources*, 140, pp. 34-40

Silva, V.S.; Schirmer, J.; Reissner, R.; Ruffmann, B.; Silva, H.; Mendes, A.; Madeira, L.M. & Nunes, S.P. (2005b). *J. Power Sources*, 140, pp. 41-49

Silva, V.S.; Ruffmann, B.; Vetter, S.; Mendes, A.; Madeira, L.M. & Nunes, S.P. (2005c). *Catalysis Today*, 104, pp. 205-212

Soczka-Guth, T.; Baurmeister, J.; Frank, G. & Knauf, R. (1999). *International Patent* WO 99/29763

Song, M.K.; Park, T.S.B.; Kim, Y.T. *et al.* (2004). *Electrochim. Acta*, 50, pp. 639-643

Staiti, P.; Aricò, A.S.; Baglio, V.; Lufrano, F.; Passalacqua, E. & Antonucci, V. (2001). *Solid State Ionics*, 145, pp. 101-107

Steck, A.E. (1995). Membrane materials in fuel cells, In: *Proceedings of the First International Symposium on New Materials for Fuel Cell Systems* 1, Savadogo, O.; Roberge, P.R. & Veziroglu, T.N. (Eds.), pp.74, Montréal, Quebec, Canada, July 9-13, 1995

Steck, A. & Stone, C. (1997). In: *Proceedings of the Second International Symposium on New Materials for Fuel Cell and Modern Battery Systems*, Savadogo, O. & Roberge, P.R. (Eds.), pp. 792-807, Montréal, Canada, July 6-10, 1997

Strickland, G. (1984). *Int. J. Hydrogen Energy*, 9, pp. 759-766

Su, Y.H.; Liu, Y.L.; Sun, Y.M.; Lai, J.Y.; Wang, D.M.; Gao, Y.; Liu, B. & Guiver, M.D. (2007). *J. Membr. Sci.*, 296, pp. 21-28

Sun, H.; Sun, G.; Wang, S.; Liu, J.; Zhao, X.; Wang, G.; Xu, H.; Hou, S. & Xin, Q. (2005). *J. Membr. Sci.*, 259, pp. 27-33

Sundholm, F. (1998). New Polymer Electrolytes for Low Temperature Fuel Cells. In: *Proceedings of the 9th International Conference on Solid State Protonic Conductors SSPC'98*, pp. 155-158, Extended Abstract Book, Bled, Slovenia, 17-21 August, 1998

Surampudi, S.; Narayanan, S.R.; Vamos, E.; Frank, H.; Halpert, G.; LaConti, A.; Kosek, J.; Surya Prakash, G.K. & Olah, G.A. (1994). *J. Power Sources*, 47, pp. 377-385

Takahashi, K.; Takezawa, N. & Kobayashi, H. (1982). *Appl. Cat.*, 2, pp. 363-366

Takeishi, K. & Suzuki, H. (2004). *Appl. Cat. A*, 260, pp. 111-117

Takezawa, N.; Kobayashi, H.; Kamegai, Y. & Shimokawabe, M. (1982). *Appl. Cat.*, 3, pp. 381-388

Tamura, K.; Tsukui, T.; Kamo, T. & Kudo, T. (1984). *Hitachi Hyoron*, 66, pp. 135-138

Tang, H.; Pintauro, P.N.; Guo, Q. & O'Connor, S. (1999). *J. Appl. Polym. Sci.*, 71, pp. 387-399

Tang, H.; Pan, M.; Jiang, S.; Wan, Z. & Yuan, R. (2005). *Colloids and Surfaces A: Physicochem. Eng. Aspects*, 262, pp. 65-70

Tay, S.W.; Zhang, X.; Liu, Z.; Hong, L. & Chan, S.H. (2008). *J. Membr. Sci.*, 321, pp. 139-145

Tazi, B. & Savadogo, O. (2000). *Electrochim. Acta*, 45, pp. 4329-4339

Tazi, B. & Savadogo, O. (2001). *J. New Mater. Electrochem. Systems*, 4, pp. 187-196

Tchicaya-Bouckary, L.; Jones, D.J. & Roziere, J. (2002). *Fuel Cells*, 2, pp. 40-45

Tian, H. & Savadogo, O. (2005). *Fuel Cells*, 5, pp. 375-382

Tricoli, V. & Nannetti, F. (2003). *Electrochim. Acta*, 48, pp. 2625-2633

Triphathi, B.P. & Shahi, V.K. (2007). *J. Colloid and Interface Science*, 316, pp. 612-621

Triphathi, B.P.; Kumar, M. & Shahi, V.K. (2009). *J. Membr. Sci.*, 327, pp. 145-154

Tung, S.P. & Hwang, B.J. (2007). *Fuel Cells*, 7, pp. 32-39

Uchida, M.; Aoyama, Y.; Eda, N. & Ohta, A. (1995). *J. Electrochem. Soc.*, 142, pp. 463-468

Ueda, S.; Eguchi, M.; Uno, K.; Tsutsumi, Y. & Ogawa, N. (2006). *Solid State Ionics*, 177, pp. 2175-2178

Uma, T. & Nogami, M. (2007). *Fuel Cells*, 7, pp. 279-284

Velu, S.; Suzuki, K. & Osaki, T. (1999). *Catal. Lett.*, 62, pp. 159-167

Verma, L.K. (2000). *J. Power Sources*, 86, pp. 464-468

Wainright, J.S.; Wang, J.-T.; Weng, D.; Savinell, R.F. & Litt, M. (1995). *J. Electrochem. Soc.*, 142, pp. L121-L123

Wainright, J.S.; Savinell, R.F. & Litt, M.H. (1997). In: *Proceedings of the Second International Symposium on New Materials for Fuel Cell and Modern Battery Systems*, Savadogo, O. & Roberge, P.R. (Eds.), pp. 808-817, Montréal, Canada, July 6-10, 1997

Walker, M.; Baurngdrtner, K.-M.; Kaiser, M. et al., (1999). *J. Appl. Polym. Sci.*, 74, pp. 67-73

Wang, J.; Wasmus, S. & Savinell, R.F. (1995). *J. Electrochem. Soc.*, 142, pp. 4218-4224

Wang, J.-T.; Savinell, R.F.; Wainright, J.; Litt, M. & Yu, H. (1996a). *Electrochim. Acta*, 41, pp. 193-197

Wang, J.-T.; Wainright, J.S.; Savinell, R.F. & Litt, M. (1996b). *J. Appl. Electrochem.*, 26, pp. 751-756

Wang, S.; Ishihara, T. & Takita, Y. (2002). *Appl. Cat. A,* 228, pp. 167-176

Wang, Y.; Yang, D.; Zheng, X.; Jiang, Z. & Li, J. (2008). *J. Power Sources,* 183, pp. 454-463

Wang, Z.H.; Wang, C.Y. & Chen, K.S. (2001). *J. Power Sources,* 94, pp. 40-50

Watanabe, M. (1995). *US Patent* 5,472,799

Watanabe, M.; Uchida, H.; Seki, Y.; Emori, M. & Stonchart, P. (1996). *J. Electrochem. Soc.,* 143, pp. 3847-3852

Watanabe, M.; Uchida, H. & Emori, M. (1998). *J. Electrochem. Soc.,* 145, pp. 1137-1141

Wei, J.; Stone, C. & Steck, A.E. (1995a). *US Patent* 5,422,411

Wei, J.; Stone, C. & Steck, A.E. (1995b). WO 95/08581, 30 March 1995

Williams, K.R. (1966). In: *An Introduction to Fuel Cells,* pp. 152, Elsevier

Wilson, M.S. & Gottesfeld, S. (1992). *J. Electrochem. Soc.,* 139, pp. L28-L30

Wilson, M.S. (1993). *US Patent* 5,211,984

Wilson, M.S.; Valerio, J.A. & Gottesfeld, S. (1995). *Electrochim. Acta,* 40, pp. 355-363

Woo, Y.; Oh, S.Y.; Kang, Y.S. & Jung, B. (2003). *J. Membr. Sci.,* 220, pp. 31-45

Wycisk, R. & Pintauro, P.N. (1996). *J. Membr. Sci.,* 119, pp. 155-160

Xing, B. & Savadogo, O. (1999). *J. New Mater. Electrochem. Systems,* 2, pp. 95-101

Xing, P.; Robertson, G.P.; Guiver, M.D.; Mikhailenko, S.D.; Wang, K. & Kaliaguine, S. (2004). *J. Membr. Sci.,* 229, pp. 95-106

Xu, K.; Lao, S.J.; Qin, H.Y.; Liu, B.H. & Li, Z.P. (2010). *J. Power Sources,* 195, pp. 5606-5609

Xu, W.; Liu, C.; Xue, X.; Su, Y.; Lv, Y.; Xing, W. & Lu, T. (2004). *Solid State Ionics,* 171, pp. 121-127

Yang, C.; Costamagna, P.; Srinivasan, S.; Benziger, J. & Bocarsly, A.B. (2001a). *J. Power Sources,* 103, pp. 1-9

Yang, C.; Srinivasan, S.; Aricò, A.S.; Cretì, P.; Baglio, V. & Antonucci, V. (2001b). *Electrochem. Solid State Lett.,* 4, pp. A31-A34

Yang, C.; Srinivasan, S.; Bocarsly, A.B.; Tulyani, S. & Benziger, J.B. (2004). *J. Membr. Sci.,* 237, pp. 145-161

Yen, S.-P.S.; Narayanan, S.R.; Halpert, G.; Graham, E. & Yavrouian, A. (1998). *US Patent* 5,795,496

Yi, J.S. & Nguyen, T.V. (1999). *J. Electrochem. Soc.,* 146, pp. 38-45

Yoon, S.R.; Hwang, G.H.; Cho, W.I.; Oh, I.H.; Hong, S.A. & Ha, H.Y. (2002). *J. Power Sources,* 106, pp. 215-223

Yoshida, N.; Ishisaki, T.; Watakabe, A. & Yoshitake, M. (1998). *Electrochim. Acta,* 43, pp. 3749-3754

Yu, R.H.; Choi, H.G. & Cho, S.M. (2005). *Electrochem. Commun.,* 7, pp. 1385-1388

Zaidi, S.M.J.; Mikhailenko, S.D.; Robertson, G.P.; Guiver, M.D. & Kaliaguine, S. (2000). *J. Membr. Sci.,* 173, pp. 17-34

Zawodzinski, T.A. Jr.; Neeman, M.; Sillerud, L.O. & Gottesfeld, S. (1991). *J. Phys. Chem.,* 95, pp. 6040-6044

Zawodzinski, T.A. Jr.; Springer, T.E.; Davey, J.; Jestel, R.; Lopez, C.; Valerio, J. & Gottesfeld, S. (1993). *J. Electrochem. Soc.,* 140, pp. 1981-1985

Zhai, Y.; Zhang, H.; Hu, J. & Yi, B. (2006). *J. Membr. Sci.,* 280, pp. 148-155

Zhang, Q.; Li, X.; Fujimoto, K. & Asami, K. (2005). *Appl. Cat. A,* 288, pp. 169-174

Zhang, W.; Tang, C.-M. & Kerres, J. (2001). *Sep. Purif. Technol.,* 22-23, pp. 209-221

Zhou, W.J.; Zhou, B.; Li, W.Z.; Zhou, Z.H.; Song, S.Q.; Sun, G.Q.; Xin, Q.; Douvartzides, S.; Goula, M. & Tsiakaras, P. (2004). *J. Power Sources,* 126, pp. 16-22

Zhou, X.; Weston, J.; Chalkova, E.; Hofmann, M.A.; Ambler, C.M.; Allcock, H.R. & Lvov, S.N. (2003). *Electrochim. Acta,* 48, pp. 2173-2180

8

Phase Diagrams in Chemical Engineering: Application to Distillation and Solvent Extraction

Christophe Coquelet[1,2] and Deresh Ramjugernath[1]
[1]Thermodynamic Research Unit,
School of Chemical Engineering,
University KwaZulu Natal,
Howard College Campus, Durban
[2]MINES ParisTech,
CEP/TEP - Centre Énergétique et Procédés, Fontainebleau
[1]South Africa
[2]France

1. Introduction

By definition, a phase diagram in physical chemistry and chemical engineering is a graphical representation showing distinct phases which are in thermodynamic equilibrium. Since these equilibrium relationships are dependent on the pressure, temperature, and composition of the system, a phase diagram provides a graphical visualization of the effects of these system variables on the equilbrium behavior between the phases. Phase diagrams are essential in the understanding and development of separation processes, especially in the choice and design of separation unit operations, e.g. knowledge about high pressure phase equilibria is essential not just in chemical processes and separation operations, but is also important for the simulation of petroleum reservoirs, the transportation of petroleum fluids, as well as in the refrigeration industry. In order to utilize the knowledge of phase behavior it is important to represent or correlate the phase information via the most accurate thermodynamic models. Thermodynamic models enable a mathematical representation of the phase diagram which ensures comprehensive and reproducible production of phase diagrams. The measurement of phase equilibrium data is necessary to develop and refine thermodynamic models, as well as to adjust them by fitting or correlating their parameters to experimental data. Generally the measurement of phase equilibria is undertaken using two categories of experimental techniques, viz. synthetic and analytic methods. The choice of the technique depends on the type of data to be determined, the range of temperatures and pressures, the precision required, and also the order of magnitude of the phase concentrations expected.

2. Definition of phases, phase transitions, and equilibrium

A phase is a homogeneous space, which can be composed of one or more components or chemical species, in which the thermodynamic properties, e.g. density or composition, are identical. A system can comprise of one or several phases. Depending on whether one or more phases exist, the system can be defined as a monophasic homogenous system in which the composition and thermodynamic properties are identical in the whole space, or as a multiphase heterogenous system for which thermodynamic properties change distinctly at the phase interface. A phase is characterized by its temperature, density, pressure, and other thermodynamics properties, i.e. Gibbs energy, molar enthalpy, entropy, heat capacity, etc.

The concept of a phase can be employed to distinguish between the different states of matter. Matter is generally accepted to exist in three main states, viz. gas, liquid, and solid. Other types of phases do exist, e.g. nematic and smectic phase transitions in liquid crystral, but this will not be relevant to this chapter. Molecular interactions between the components or chemical species which comprise the system are responsible for the different states of matter. Consequently thermodynamics model used to describe phases have to take into account the molecular interactions, or be used at conditions at which they are relevant, e.g. the ideal gas model can only be applied to gas phase at low pressures and sufficiently high temperatures, as at these conditions molecular interactions are negligible.

2.1 Phases transitions

There a number of definitions that can be used to describe a liquid. One of the definitions, which is easy to visualize but may not be entirely thermodynamically correct, is that a liquid is a fluid which takes the form of its containment without necessarily filling the entire containment volume. This characteristic distinguishes a liquid from a solid, as well as a liquid from a gas. As stated, it is not an accurate definition and it is necessary to carefully check the thermophysical microscopic (structure) and macroscopic properties of the state of matter.

A phase change is the transition between different states of matter. It is generally characterized by a sudden change of the internal microscopic structure and the macroscopic properties of the environment. It is probably better to refer to it as a phase transition instead of phase change because a phase transition doesn't imply the change of state of the matter, e.g. liquid-liquid or solid-solid phase transitions. For a solid-solid phase transition there can be a change of the structure of the crystal.

A phase transition can be effected by a change of the composition, temperatures and/or pressures of the system, or by application of an external force on the system. Consequently, composition, density, molar internal energy, enthalpy, entropy, refractive index, and dielectric constant have different values in each phase. However, temperature and pressure are identical for all phases in multiphase systems in compliance with thermodynamic principles. As a result, when two phases (or more) exist, we refer to it as phase equilibria. Table 1 lists all the phase transitions between solid, liquid and vapor phases.

Phase 1	Phase 2	Transition 1-2
liquid	vapor	boiling
liquid	solid	solidification
vapor	liquid	liquefaction
vapor	solid	condensation
solid	vapor	sublimation
solid	liquid	melting

Table 1. Phases transitions.

There are two types of phase transition:

First order: In this type of transition there is a discontinuity in the first derivative of the Gibbs free energy with regard to thermodynamic variables. For this type of transition there is an absorption or release of a fixed amount of energy with the temperature remaining constant.

Second order: For this type of transition, there is continuity in the first derivative of the Gibbs free energy with regard to the thermodynamic variable across the transition, but there is a discontinuity in the second derivative of the Gibbs energy with regard to thermodynamic properties.

2.2 Chemical potentials and equilibrium conditions

2.2.1 Chemical potential

The chemical potential is the one of the most important thermodynamic properties used in the description of phase equilibrium. If one considers a phase with volume V, containing n_c components at temperature T and pressure P, the chemical potential μ_i of component i in the phase is defined by

$$\mu_i\left(P,T,n_1,n_2,...,n_C\right)=\left(\frac{\partial G\left(P,T,n_1,n_2,...,n_C\right)}{\partial n_i}\right)_{T,P,n_{j\neq i}} \tag{1}$$

Where G=H-TS is the Gibbs free energy of the phase. The expression for the infinitesimal reversible change in the Gibbs free energy is given by

$$dG = VdP - SdT + \sum_i \mu_i dn_i \tag{2}$$

Moreover, from Euler theorem we can write $G = \sum_{i=1}^{N_c} n_i \mu_i$. The Gibbs-Duhem (Eq. 3) equation can be obtained from equations 1 and 2:

$$VdP - SdT + \sum_i n_i d\mu_i = 0 \tag{3}$$

2.2.2 Equilibrium conditions

Considering a multicomponent system in equilibrium between two phases (α and β), at temperature T and pressure P, the Gibbs free energy of this system is $G=G^\alpha+G^\beta$, where G^α and G^β, are the Gibbs free energy of the α and β phases respectively. At equilibrium, the Gibbs free energy must be at the minimum, therefore ($dG=0$):

$$dG = dG^\alpha + dG^\beta = \sum_i \mu_i^\alpha dn_i^\alpha + \sum_i \mu_i^\beta dn_i^\beta = 0 \tag{4}$$

For a closed system, $dn_i^\beta = -dn_i^\alpha$ for each component i between phases α and β. For each chemical species, the following relationship can written:

$$\mu_i^\alpha\left(P,T,n_1^\alpha,n_2^\alpha,...,n_{N_C}^\alpha\right)=\mu_i^\beta\left(P,T,n_1^\beta,n_2^\beta,...,n_{N_C}^\beta\right) \tag{5}$$

Consequently, the equilibrium condition a of multiphase mixture is equality of temperature, pressure, and chemical potential μ_i of each component i in all the phases in equilibrium.

3. Description of phase diagrams

3.1 Pure compound

3.1.1 Description

The phase diagram of a pure compound is characterized by its critical point, triple point, vapor pressure, and melting and sublimation curves (figure 1). At the triple point, the solid, liquid, and vapor phases coexist. The critical point can be defined as the upper limit for the pure component vapor pressure curve. For temperatures and pressures above the critical conditions, there is no possibility to have vapor-liquid equilibrium. The supercritical state can be considered as a "stable state" with no possibility for phase separation.

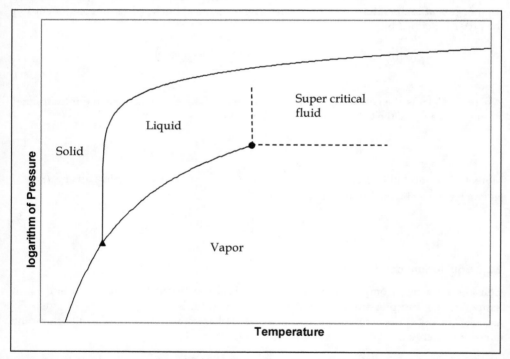

Fig. 1. Pure compound Pressure-Temperature phase diagram. (▲) Triple point, (●) : critical point. Lines : Coexistence curves

The number of intensive variables which have to be specified in order to characterize the system is determined with the Gibbs phases rule (Eq. 6).

$$F = C + 2 - \phi \tag{6}$$

Were, F, is the of degrees of freedom, C, is the number of components and Φ, is the number of phases present. Table 2 illustrates the degrees of freedom for a single component system.

Region of the phase diagram	Number of phases	Degrees of freedom	Variables to be specified
vapor pressure, or melting or sublimation curves	2	1	T or P
liquid, vapor, or solid	1	2	P and T
Triple point	3	0	Everything is fixed
Critical point	2	1	T_C or P_C

Table 2. Degrees of freedom for a pure component ($C=1$)

The phase behavior of a pure component can be represented in a plot of pressure versus temperature. Assuming that there are only two co-exisiting phases, i.e. a liquid and vapor phase, the vapor-liquid equilibrium can be represented as shown in Figure 2.

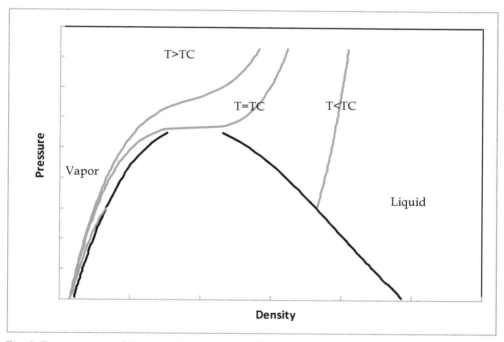

Fig. 2. Pure compound Pressure-Temperature-density phase diagram. Red line: isothermal curve. Black line saturation curve (bubble and dew points lines)

3.1.2 The critical point

The critical point of a pure compound is the upper limit (temperature and pressure) of the pure component vapor pressure curve. For temperatures and pressures below the critical point, phase separation can occur leading to two phases in equilibrium (liquid and vapour). The critical point can be considered as a limit of stability of the supercritical phase.

Critical properties (T_C, P_C, ...) characterize the critical point and the value of the properties are particularly influenced by molecular interactions, i.e. if there are strong attractive interaction between the molecules, the value of the critical temperatures and pressures are greatly increased. Using the example of water [NIST, 2010 #167] (T_C= 647.13 K, P_C= 22.055 MPa) one can clearly observe that the values of the critical properties are far greater when compared to those of methane [NIST, 2010 #167] (T_C= 190.564 K, P_C= 4.599 MPa). Hydrogen bonding between molecules of water is responsible for association effects between water molecules.

At the critical point, due to the fluctuation of the density, there is a disturbance of light waves across the visual spectrum. This phenomenon is called critical opalescence. At the critical point there is a second order type phase transition. In the Pressue-density diagram (figure 2), the isotherm labeled T_C has an inflection point and is characterized by the following relations:

$$P > 0 \text{ , } \left(\frac{\partial P}{\partial v} \right)_T = 0 \text{ , } \left(\frac{\partial^2 P}{\partial v^2} \right)_T = 0 \text{ , } \left(\frac{\partial^3 P}{\partial v^3} \right)_T < 0 \tag{7}$$

Experimentally, it can be observed that along the coexisting curve liquid and vapor densities diverge at the critical point following the relationship $\rho_L - \rho_V \propto (T_C - T)^\beta$ where ρ_C and T_C are the critical density and critical temperature respectively and β is an universal critical exponent (value is around 0.326, whatever the pure compound). This mathematical relationship is one of thermodynamics relationships and referred to as scaling laws. For pressure, one can write the following relationship along the critical isotherm curve:

$\left| \dfrac{P - P_C}{P_C} \right| \propto \left| \dfrac{\rho - \rho_C}{\rho_C} \right|^\delta$, where $\delta = 4.800$. With regard to heat capacity, the following equation

can be written along the critical isochor: $C_V \propto k_{\alpha 0} + k_{\alpha 1} \left(\Delta \tau \right)^{-\alpha}$ for $\rho = \rho_C$ with $\Delta \tau = \left| \dfrac{T - T_C}{T_C} \right|$.

For isothermal compressibility the relationship along the critical isochors is $\chi \propto k_{\alpha 1} \left(\Delta \tau \right)^{-\gamma}$

for $\rho = \rho_C$ with $\Delta \tau = \left| \dfrac{T - T_C}{T_C} \right|$. The exponents' δ, α and γ are also referred to as critical

exponents. Simple relationships can be written between these exponents, e.g. $\alpha + 2\beta + \gamma = 2$. All of these laws are also observed experimentally and can be derived from renormalisation theory. Table 3 shows the optimum values for these critical exponents.

Critical exponent	Values
α	0.110
β	0.3255
γ	1.239
δ	4.800

Table 3. Critical exponents

3.1.3 Triple point

At the triple point, the three states of the matter (solid, liquid, and vapor) are in equilibrium. The temperature and pressure of the triple point is fixed because the degrees of freedom at this point is equal to zero. Triple point values are very useful for the definition of reference points on temperature scales and for the calibration of experimental equipment. The table 4 lists triple point conditions for some for some common chemicals.

Compound	T_{tr}/K	P_{tr}/kPa
Water	273.16	0.6117
Ethanol	150	4.3×10^{-7}
Oxygen	54.36	0.152
Methane	90.67	11.69

Table 4. Triple points of few pure compounds (Ref. : NIST [NIST, 2010 #167]).

3.2 Binary systems

The first equation of state, which could describe the behaviour of both liquid and vapour states of a pure component, was developed by (van der Waals, 1873). Two types of interactions (repulsive and attractive) were considered in this equation.

$$\left(P + \frac{a}{v^2}\right)(v - b) = RT \tag{8}$$

The stability criteria are defined by the following equations:

$$\left(\frac{\partial P}{\partial v}\right)_T < 0, \left(\frac{\partial^2 P}{\partial v^2}\right)_T < 0 \text{ and } \frac{T}{c_v} > 0$$

At the critical point these two previous conditions are equal to zero.

$$\left(\frac{\partial P}{\partial v}\right)_T = 0, \left(\frac{\partial^2 P}{\partial v^2}\right)_T = 0 \text{ and } \frac{T}{c_v} = 0$$

It can be seen that at the critical point, the isochoric heat capacity has infinite value. Phase diagram describing binary mixtures depend of the behaviour of the species. Van Konynenburg & Scott (1980) have classified the phase behaviour of binary mixtures into six types considering van der Waals EoS and quadratic mixing rules. Figure 3 presents the different types of phase diagrams. The transition between each type of phase diagram can be explained by considering the size effects of molecules and the repulsive interactions between them. Figure 4 illustrates the possible transitions between the different types of phase diagrams.

3.2.1 Mixture critical point

The thermodynamic stability of a mixture determines if it would remain as a stable homogenous fluid or split into more stable phases and therefore produce two or more

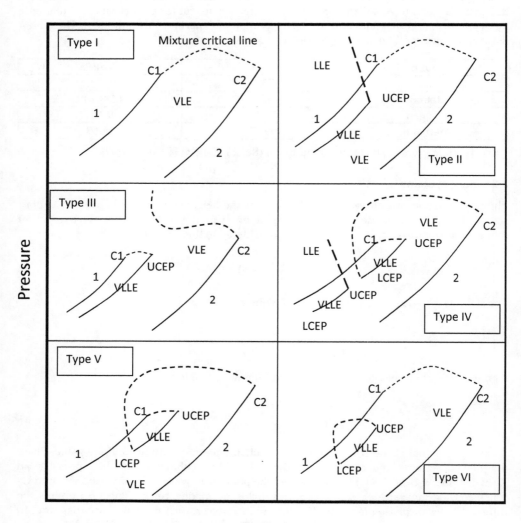

Fig. 3. Six types of phase behaviour in binary fluid systems. C: Critical point, L: Liquid, V: Vapor. UCEP: Upper critical end point, LCEP: Lower critical end point. Dashed curve are critical.

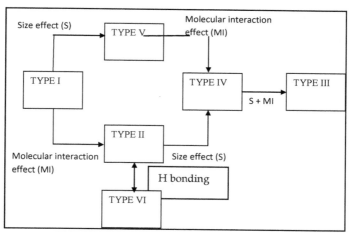

Fig. 4. Evolution of phase diagrams.

phases in equilibrium. By definition, a mixture is considered as stable when the Gibbs or Helmholtz free energy are at their minimum. Figure 5 gives an indication of the change of the Gibbs free energy with composition for a given temperature and pressure from stable condition (a) (T, P) to an unstable condition (b). In figure 5 b, one can observe that there are

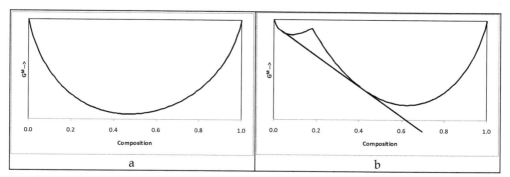

a): stable condition, b): unstable condition with phase equilibria.

Fig. 5. Mixing Gibbs free energy (G) as a function of molar composition (x1) at given T and P.

two minima: this corresponds to phase equilibria. More details on thermodynamic phase stability are given by (Michelsen, 1982a). A Taylor series expansion can be obtained for the Gibbs free energy for a given temperature:

$$G = \sum_i n_i^0 \mu_i^0 + \sum_i \mu_i^0 \Delta n_i + \frac{1}{2}\sum_i \sum_j \left(\frac{\partial^2 G}{\partial n_i \partial n_j}\right)_{T,P} \Delta n_i \Delta n_j +$$

$$\frac{1}{6}\sum_i \sum_j \sum_k \left(\frac{\partial^3 G}{\partial n_i \partial n_j \partial n_k}\right)_{T,P} \Delta n_i \Delta n_j \Delta n_k + \Theta\left(\Delta n^5\right)$$

(9)

With $n_i = n_i^0 + \Delta n_i$ and $V = V^0 + \Delta V$.

The stability condition leads to:

$$G - \sum_i n_i^0 \mu_i^0 - \sum_i \mu_i^0 \Delta n_i \geq 0 \qquad (10)$$

Consequently

$$\frac{1}{2}\sum_i\sum_j\left(\frac{\partial^2 G}{\partial n_i \partial n_j}\right)_{T,P}\Delta n_i \Delta n_j + \frac{1}{6}\sum_i\sum_j\sum_k\left(\frac{\partial^3 G}{\partial n_i \partial n_j \partial n_k}\right)_{T,P}\Delta n_i \Delta n_j \Delta n_k + \Theta\left(\Delta n^5\right) \geq 0 \qquad (11)$$

The critical point can be described as the limit of stability (x is the molar fraction) and coordinates (P, v) can be determined considering the following relations. More details are given by (Michelsen, 1980, 1982b), (Heidmann & Khalil, 1998) and (Stockfleth & Dohrn,1980).

$$G_{2x} = \left(\frac{\partial^2 G}{\partial n_i \partial n_j}\right)_{T,P} = 0 \; ; \; G_{3x} = \left(\frac{\partial^3 G}{\partial n_i \partial n_j \partial n_k}\right)_{T,P} = 0 \; ; \; G_{4x} = \left(\frac{\partial^4 G}{\partial n_i \partial n_j \partial n_k \partial n_l}\right)_T > 0 \qquad (11)$$

In their paper, (Baker et al., 1982) present several examples of Gibbs energy analysis. Figure 6 compares phase diagrams (pressure/composition) and the Mixing Gibbs free energy at a given temperature.

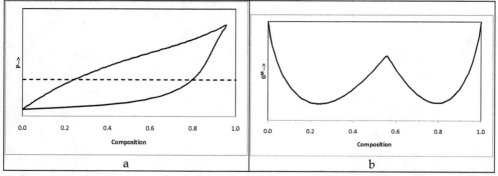

A) Phase diagram at a given temperature. Dashed line: given pressure. B) Gibbs free energy of mixing at the same temperature and given pressure.

Fig. 6. Example of Mixing Gibbs free energy minimum for a binary system.

3.2.2 Phase diagram classification

Type I phase behaviour

It is the simplest type of phase diagram. The mixture critical point line starts at the first pure component critical point and finish at the second pure component critical point. Mixtures which exhibit this behaviour are two components which are chemically similar or have comparable critical properties., e.g. systems with CO_2 and light hydrocarbons, systems with HFC refrigerants, and benzene + toluene binary systems.

Type II phase behaviour

It is similar to type I but at low temperatures, the two components are not miscible in the liquid mixtures. Consequently a liquid - liquid equilibrium appears. The mixture critical point line for liquid - liquid equilibrium starts from the UCEP (upper critical end point). At the UCEP, the two liquid phases merge into one liquid phase. Examples of systems which exhibit this behaviour are those with hydrocarbons and fluorinated fluorocarbons.

Type III phase behaviour

Generally system which have very large immiscibility gaps exhibit this behaviour, e.g. aqueous systems with hydrocarbons. A liquid – liquid - vapour curve appears and a first mixture critical point line starts from the pure component 1 critical point and ends at the UCEP. The second one starts from the infinite pressure ($P \rightarrow \infty$) and ends at the pure component 2 critical point, generally the solvent i.e. water. The slope of this second curve can be positive, negative or positive and negative. Concerning the positive curve, we have two phases at temperatures larger than the critical temperature of pure component two.

Some examples:

Positive slope: helium + water binary system

Negative slope: methane + toluene binary system

Positive and negative slope: nitrogen + ammonia or ethane + methanol binary systems.

Type IV phase behaviour

It is similar to type V behaviour. The vapor-liquid critical point line starts at the critical point of component 2 and ends at the LCEP (lower critical end point). Vapour-liquid-liquid equilibrium (VLLE) exists and is present in two parts. Ethane + n-propanol and CO_2 + nitrobenzene binary systems are examples of systems which show this behaviour.

Type V phase behaviour

It is a modification of type III phase diagram. There are two vapor-liquid critical point lines. One goes from the pure component critical point 1 and ends at the UCEP. The other starts at the pure component critical point 2 and ends at the LCEP. Contrary to type IV systems, below the LCEP the liquids are completely miscible. The ethylene + methanol binary system is an example of a type V system.

Type VI phase behaviour

There are two critical point curves. The first is similar to one presented with the type II diagram: a connection between the two pure component critical points. The second connects the LCEP and the UCEP. Between these two points, there exists VLLE. The system "water + 2-butanol" is a typical example of a type VI system. In fact the main reason is due to the existence of hydrogen bonding for one or both of the pure components (self association) and in the mixture (strong H bonding between the two components). H bonding favours the heat of solution and so miscibility in the liquid state. Above the LCEP hydrogen bonds break and the liquid becomes unstable and a second liquid phase appears.

3.3 Ternary systems

There exists a number of different classifications for ternary systems, but we propose that of Weinstsock (1952). According to this classification the system can exhibit vapor-liquid, liquid-liquid, and also vapor-liquid-liquid equilibrium (VLE, LLE and VLLE). Considering vapor-liquid-liquid equilibria, the phase diagrams can be classified into 3 categories or types. For ternary diagrams, in general, temperature and pressure are fixed.

Type 1 :

It is the most common diagram. In figure 7, one can observe VLE and the monophasic regions. The shape of the phase diagram changes with the pressure. In figure 8, one can see the phase diagram for the ternary system comprising R32 + R227ea + R290 (Coquelet et al., 2004).

Type 2 :

It is an evolution of a type 1 phase diagram: VLLE and LLE appear but the two liquids are partially miscible (see figure 9).

Type 3 :

According to figure 10, there exists strong immiscibility between two species. The size of the LLE region increases with pressure. Consequently, LLE region disappears to form a Liquid Fluid Equilibrium region (if there is a supercritical fluid).

There is also classification for Liquid – liquid equilibrium phase diagram. There are generally three distinguishable categories (Figure 11) :

-   One pair of binary components are partially miscible (type 1)
-   Two pairs of binary components are partially miscible (type 2)
-   All three pairs of binary components are partially miscible (type 3)

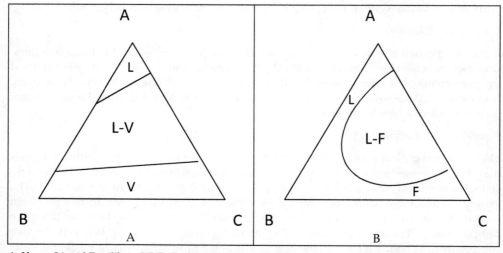

A: Vapor-Liquid Equilibria (VLE), B: if T or P is increased a supercrital fluid (F) appears and so a FLE.

Fig. 7. Ternary diagram: type 1.

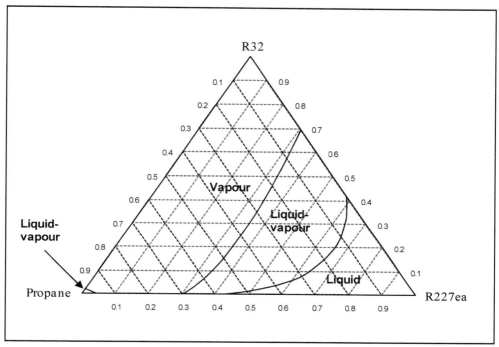

Fig. 8. Phase diagram of the ternary system R32-R290-R227ea at T=293 K and P=8.5 bar (Coquelet et al., 2004).

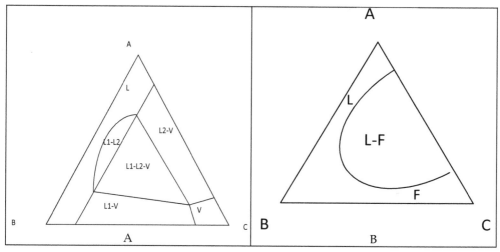

Fig. 9. Ternary diagram: type 2. A: Existence of Vapor Liquid Liquid Equilibrium, B: if we increase pressure or temperature, we have a supercritical fluid in equilibrium with one liquid phase (FLE).

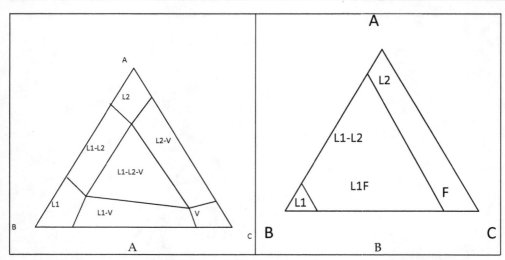

Fig. 10. Ternary diagram: type 3. A: There exists Liquid-Liquid Equilibria (L1L2E) between species A and B: if we increase pressure or temperature, we have a supercritical fluid in equilibrium with one liquid phase (FLE) and L1L2E.

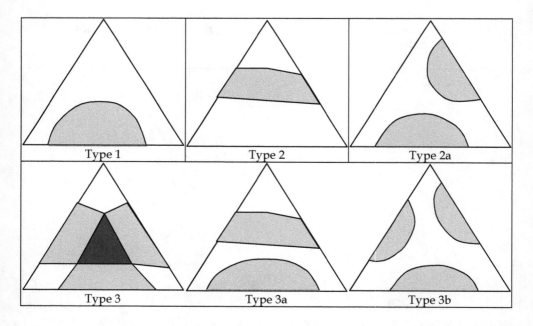

Fig. 11. Liquid – liquid equilibrium : presentation of the 3 configurations of the type 1, 2 and 3 systems.

3.4 Pressure-temperature envelops

The Pressure-Temperature (P-T) envelop is a very interesting way to represent phase diagrams. For a mixture (where composition is known), the P-T envelop represents the limits of the phase equilibrium region. An example of a P-T envelop is illustrated in figure 12. The point which corresponds to the maximum of pressure is called the cricondenbar, and with regard to the maximum of temperature, it is called the cricondentherm. Bubble and dew pressures curves are also presented in such a diagram with the critical point. With such a diagram, phenomenon such as retrograde condensation can be easily explained.

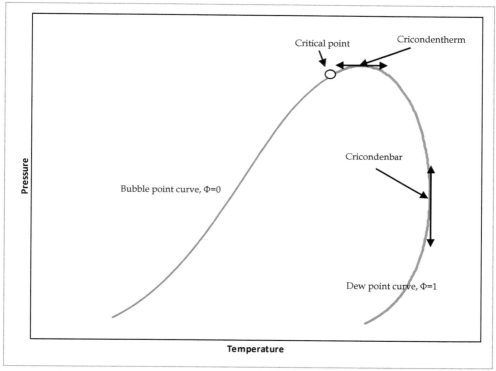

Fig. 12. Example of P-T envelop of hydrocarbons (C1 to C5).

4. Application to distillation

Distillation is the most well known separation unit operation in the world. Utilizing energy, the objective is to create one or more coexisting zones which differ in temperature, pressure, composition and phase state. In order to design a distillation column, the concept of an equilibrium stage is required: at this stage, the vapor and liquid streams which are leaving the stage are in complete equilibrium with each other and thermodynamic relations can be used to determine the temperature and the concentration of the different species for a given pressure. The equilibrium stage can be simulated as a thermodynamic isothermal FLASH (Michelsen, 1982b). Consequently, for a given distillation column, if the number of equilibrium stages is very large it means that the separation is difficult due a relative volatility close to one. The relative volatility of component i with respect to component j is the ratio between the partition coefficient (or equilibrium constant) (Eq. 12).

$$\alpha_{ij} = \frac{K_i}{K_j} = \frac{y_i/x_i}{y_j/x_j}$$ (12)

In distillation, three types of binary equilibrium curves are shown in figure 13.

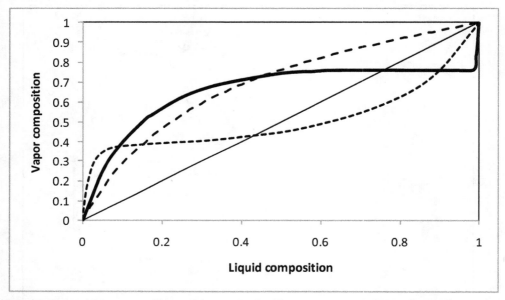

Fig. 13. Typical binary equilibrium curves. Dashed line: system with normal volatility, dotted line: system with homogenous azeotrope and solid line: system with heterogeneous azeotrope

If the relative volatility is equal to one, it is impossible to separate the two components. The phase diagrams are very important for the design of distillation columns: considering the McCabe and Thiele method (Perry & Green, 1997), the knowledge of the y-x curve together with the stream flowrates lead to the determination of the number of theorical equilibrium stage (NTES). The McCabe-Thiele method is based on the assumption of constant molar overflow and the molar heats of vaporization of the feed components being equal; heat effects such as heats of mixing, heat transfer to and from the distillation column are negligible. If the NTES is very large, the separation is difficult. Figure 15 shows phase diagrams for an air separation unit. It can clearly be seen that the separation between O_2 and Ar is more difficult in comparison with O_2 and N_2.

Considering a distillation column with a total reflux, the closer the values of liquid and vapor compositions, the higher is the number of equilibrium stages and more difficult is the

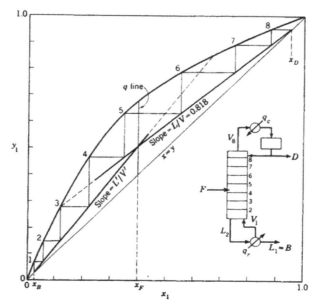

Fig. 14. Example of McCabe and Thiele construction: system Extract from Perry's handbook (Perry & Green, 1997).

phase separation. For a homogeneous azeotrope, the relative volatility is equal to one. The definition of the existence of an azeotrope is as following: $\left(\dfrac{\partial P}{\partial x_i}\right)_T = 0$ or $\left(\dfrac{\partial T}{\partial x_i}\right)_P = 0$ or $y_i = x_i$ for the Nth component. Moreover, considering only VLE, the mathematical criterion can determine whether there existences an azeotrope in a system which presents VLLE behavior: in this case there exists a heterogeneous azeotrope. Table 5 describes the different types or categories of azeotropes with example [12].

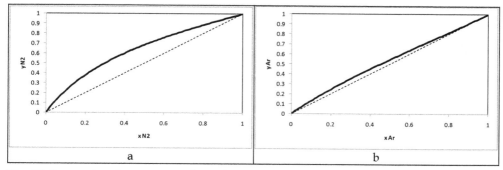

Fig. 15. Comparison between y-x phase diagram of N_2-O_2 (a) and Ar-O_2 (b) binary systems at 110 K.

Category	Type of azeotrope	Examples
I	Homogeneous Maximum of pressure maximale	1-propanol + water ethanol + benzene
II	Heterogeneous azeotrope Maximum of pressure maximale	1-butanol + water water + benzene
III	Homogeneous Minimum of pressure	Trichloromethane + 2-butanone
IV	Homogeneous Maximum of pressure with immiscibility zone	2-butanone + water 2-butanol + water
V	Double azeotrope Local minimum and maximum of pressure	benzene + hexafluorobenzene Diethylamine + methanol
VI	Homogeneous Minimum of pressure with immiscibility zone	Triethylamine + acetic acid

Table 5. Classification of azeotropes (fixed temperature).

Figure 16 presents an example of a heterogeneous azeotropic system (2 butanol + water) at 320 K.

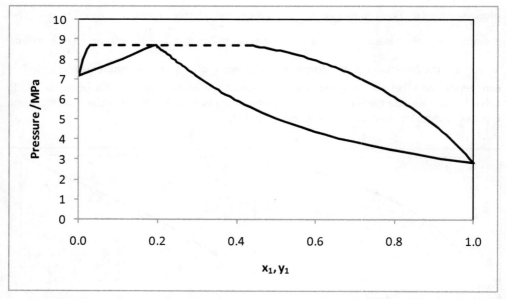

Fig. 16. 2-butanol (1) + water(2) binary system. Prediction using PSRK EoS (Holderbaum & Gmehling, 1991) at 320 K. Dashed line: Vapor Liquid Liquid equilibria (Heterogenous azeotrope)

5. Application to liquid-liquid extraction

Liquid-liquid extraction, which is commonly referred to as solvent extraction, involves the separation of the components that constitute a liquid stream by contacting it with another liquid stream which may be insoluble or partially soluble. Due to some of the components being preferentially more soluble in one of the liquid streams, separation can be effected. The separation effected in a single contacting stage is usually not large and therefore multiple contacting stages are needed to produce a significant separation. In these extraction processes the feed stream (which contains components that are to be separated) is contacted with a solvent stream. Exiting the contacting unit would be solvent-rich stream which is generally referred to as the extract and a residual liquid which is commonly called the raffinate. In general, this contacting removes a solute from the feed stream and concentrates it in the solvent-rich stream, i.e. decreasing the concentration of that particular solute in the raffinate stream.

The typical triangular diagrams which are used to illustrate ternary liquid-liquid equilibria can be converted into more convenient diagrams for visualization and computations in solvent extraction, e.g the distribution diagram, as seen in figure 17. This is undertaken because the phase relationships are generally very difficult to express conveniently algebraically and as a result solvent extraction computations are usually made graphically. The triangular diagram can also be transformed into rectangular coordinates, as seen in Figure 18.

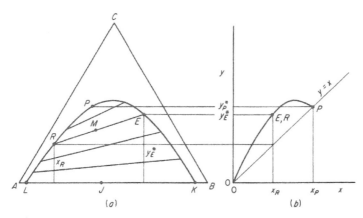

Fig. 17. Triangular (a) and distribution diagram (b) for liquid-liquid equilibria for a system of three components where one pair is partially miscible (extracted from Treybal, 1981).

The principles of separation and utilization of phase diagrams for the sizing of unit operation in solvent extraction is analogous to that which is seen in distillation, e.g. figure 18 illustrates phase diagrams being used to determine the number of theoretical stages for separation. Just as in distillation, the process can be undertaken with reflux. Reflux at the extract end can produce a product which is greater in composition, as is the case in the rectification section of a distillation column. The concept of the operating lines in the diagram, as well as the "stepping-off" in the diagram to determine the number of theoretical stages is similar to the McCabe-Thiele method for distillation.

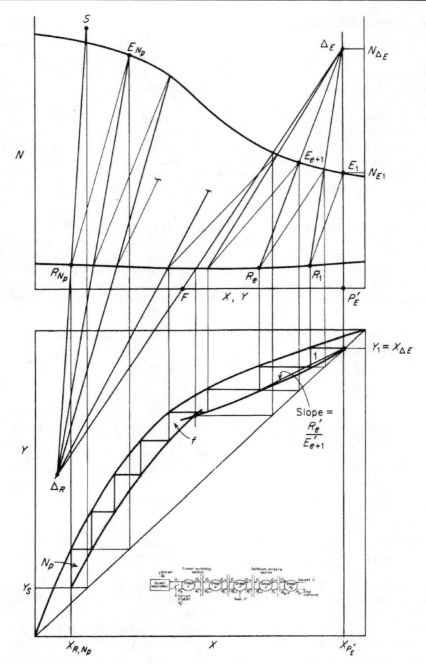

Fig. 18. Illustration of phase diagrams generated for a countercurrent extraction process with reflux (extracted from Treybal, 1981)

As the phase behaviour is affected by the choice of the solvent, and it is highly unlikely that any particular liquid will exhibit all of the properties desirable for solvent extraction, the final choice of a solvent is in most cases a compromise between various properties and parameters, viz. selectivity, distribution coefficient, insolunbility of the solvent, recoverability, density, interfacial tension, chemical reactivity, viscosity, vapor pressure, freezing point, toxicity, flammability, and cost.

Generally, the key parameters (selectivity and distribution coefficient) for determining the best solvent for the separation are calculated from liquid-liquid equilibrium measurements for the system of components concerned (including the solvent).

The selectivity is defined as follows:

$$\beta = \frac{\text{wt fraction of solute in raffinate}}{\text{wt fraction of solute in extract}} \tag{13}$$

Analogous to the relative volatility in distillation, the selectivity must exceed unity (the greater the value away from unity the better) for separation to take place. If the selectivity is unity, no separation is possible. The distribution coefficient (which is effectively the inverse of selectivity) however is not required to be larger than unity, but the larger the distribution coefficient the smaller the amount of solvent which will be required for the extraction.

Excellent summaries of solvent extraction processes are given in Perry and Green (1997), as well as in Treybal (1981) and McCabe et al. (2005).

6. Conclusion

In chemical engineering, the knowledge of the phase behaviour is very important, as the design and the optimization of the separation processes needs a good knowledge of the phase diagrams. Practically, the determination of phase diagrams can be obtained through experimental methods and/or modelling. The readers can refer to the books of Raal and Mühlauber (1998) to have a complete description of experimental methods, and Prausnitz et al. (1999) regarding models and principles. The complexity of phase diagrams is increased drastically if a solid or a polymer phase exists, however the purpose of this chapter was to introduce the reader to phase diagrams encountered by chemical engineers in the most commonly used units operations, viz. distillation and solvent extraction.

7. References

Baker, L.E., Pierce, A.C., Luks, K.D. (1982) Gibbs Energy Analysis of Phase Equilibria, *Society of Petroleum Engineers Journal*, October, pp 731-742, ISSN: 0197-7520

Coquelet, C., Chareton, A., Richon, D. (2004) Vapour–liquid equilibrium measurements and correlation of the difluoromethane (R32) + propane (R290) + 1,1,1,2,3,3,3-heptafluoropropane (R227ea) ternary mixture at temperatures from 269.85 to 328.35 K. *Fluid Phase Equilibria*, Vol. 218, pp 209-214, ISSN 0378-3812

Gmehling, J., Menke, J., Krafczyk, J., Fischer, K. (1994) *Azeotropic Data Part I*, Wiley-VCH, ISBN : 3-527-28671-3, Weinheim, Germany.

Heidemann, R. A. & Khalil, A. M. (1980) The Calculation of Critical Points, *American Institute of Chemical Engineers Journal*, Vol. 26, No.5, pp 769-779, ISSN (printed): 0001-1541. ISSN (electronic): 1547-5905

Holderbaum, T. & Gmehling, J. (1991) PSRK : A group contribution equation of state based on UNIFAC, Fluid Phase Equilibria, Vol. 70, pp 251-265, ISSN 0378-3812

McCabe, W., Smith J., Harriot, P. (2005) Unit Operations of Chemical Engineering 7th edition, McGraw-Hill Book Company, ISBN 0072848235.

Michelsen, M. L. (1980) Calculation of phase envelopes and critical point for multicomponent mixtures, *Fluid Phase Equilibria*, Vol. 4, pp 1-10, ISSN 0378-3812

Michelsen, M. L. (1982a) The isothermal Flash Problem. Part I Stability. *Fluid Phase Equilibria*, Vol. 9, pp 1-19, ISSN 0378-3812

Michelsen, M. L. (1982b) The isothermal Flash Problem. Part II Phase Split Calculation, *Fluid Phase Equilibria*, Vol. 9, pp 21-40, ISSN 0378-3812

Perry, R.H. & Green, D.W. (1997) Perry's chemical Handbook 7th , edition McGraw Hill Companies, ISBN0-07-049841-5, New York, USAPrausnitz, J.M., Lichtenthaler R.N., de Azevedo, E. G. (1999) Molecular thermodynamics of fluid phase equilbria, 3rd edition, Prenticd Hall International Series, Upper Saddle River, USA.

Raal, J. D. & Mühlbauer A. L. (1997) Phase Equilibria, Measurement and Computation, Taylor & Francis, ISBN 1-56032-550-X, London, UK.

Stockfleth, R. & Dohrn, R. (1998) An algorithm for calculating critical points in multicomponent mixtures which can be easily implemented in existing programs to calculate phase equilibria. *Fluid Phase Equilibria*, Vol. 145, pp 43-52, ISSN 0378-3812

Treybal, R.E. (1981) Mass-Transfer Operations 3rd edition, McGraw-Hill Book Company, ISBN-0-07-065176-0, Singapore.

Van der Waals, J.D. (1899) *Over de Continuiteit van den Gas- en Vloestoftoestand. (Über die Kontinuittät des Gas- und Flüssigkeitszustands) 1873*, Dissertation, Universität Leiden, Niederlande, deutsche Übersetzung, Leipzig, Germany.

Van Konynenburg, P. H. & Scott, R.L. (1980) Critical lines and Phase Equilibria in Binary van der Waals mixtures. *Philosophical Transactions of the Royal Society of London*, Vol. 298, pp 495-539, ISSN 0264-3820

Weintsock, J. J. (1952) *Phase equilibrium at elevated pressure in ternary systems of ethylene and water and organic liquids*, Phd dissertation, Princeton University, USA

9

Membrane Operations
for Industrial Applications

Maria Giovanna Buonomenna[1,*], Giovanni Golemme[1,*] and Enrico Perrotta[2]
*[1]Department of Chemical Engineering and Materials and
INSTM Consortium, University of Calabria, Rende (CS)
[2]Department of Ecology, University of Calabria, Rende (CS)
Italy*

1. Introduction

A resource-intensive industrial development, particularly in some Asian countries, characterized the last century. Its main causes can be ascribed to the significant elongation of life expectation, and to the overall increase in the standards characterizing the quality of life. The drawback of these positive aspects is the emergence of problems related to the industrial development: water stress, the environmental pollution, and the increase of CO_2 emissions into the atmosphere.

The need to achieve a knowledge-intensive industrial development is nowadays well recognized. This will permit the transition from an industrial system based on quantity to one based on quality in the framework of a sustainable development.

Sustainable development is a development that meets the needs of the present without compromising the ability of future generations to meet their own needs (United Nations General Assembly, 1987). The "three pillars" of sustainability are the environmental, social and economic demands – (United Nations General Assembly, 1987), which are not mutually exclusive and can be mutually reinforced (Figure 1): both economy and society are constrained by environmental limits. Sustainability is the path of continuous improvement, wherein the products and services required by society are delivered with progressively less negative impacts upon the Earth. Figure 2 is a cartoon depicting the road to sustainability. To benchmark sustainability, the Sustainability Index (SI) accounts for key factors that are fundamental to the industrial process (Cobb *et al*, 2007)

- strategic commitment to sustainability
- safety performance
- environmental performance
- social responsibility
- product stewardship
- innovation
- value-chain management

---

* Corresponding Authors

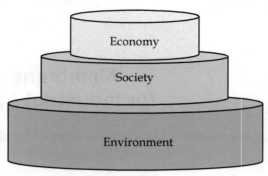

Fig. 1. Relationship between the three pillars of sustainability: environmental limits constrain both society and economy.

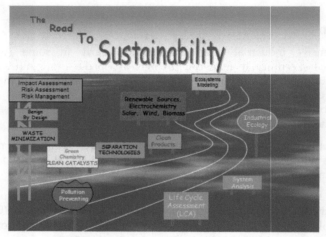

Fig. 2. Cartoon showing the features and the requirements of the road to sustainability

Chemical Engineering faces today the crucial challenge of sustainable growth to find solutions to the increasing demand for raw materials, energy and tailor-made products.

In this context, the rational integration and implementation of innovative technologies, able to increase process performance, save energy, reduce costs, and minimize the environment impact represent interesting answers.

Recently, the logic of process intensification has been suggested as the best process engineering answer to the situation. It consists of innovative equipment, design, and process development methods that are expected to bring substantial improvements in chemical and any other manufacturing and processing, such as decreasing production costs, equipment size, energy consumption, and waste generation, and improving remote control, information fluxes, and process flexibility (Charpentier, 2007).

Membrane operations are, in principle, the most attractive candidates to satisfy the process intensification concepts and requirements.

Their intrinsic characteristics of efficiency and operational simplicity, high selectivity and permeability for the transport of specific components, compatibility between different membrane operations in integrated systems, low energetic requirement, good stability under operating conditions and environmental compatibility, easy control and scale-up, and large operational flexibility, represent an interesting answer for the rationalization of industrial productions (Drioli & Romano, 2001).

The traditional membrane separation operations such as reverse osmosis (RO), microfiltration (MF), ultrafiltration (UF), and nanofiltration (NF), electrodialysis, pervaporation, etc. (Table 1), are largely used in many different applications.

Driving Force	Pressure difference (ΔP)	Concentration difference (ΔC)	Temperature difference (ΔT)	Electrical potential difference (Δf)
Phenomenological Equation	Darcy's law	Fick's law	Fourier's law	Ohm's law
Membrane Operations	Microfiltration (MF)	Gas separation (GS)	Membrane distillation (MD)	Electro-dialysis (ED)
	Ultrafiltration (UF)	Pervaporation (PV)		Electro-osmosis (EO)
	Nanofiltration (NF)	Dialysis (D)		
	Reverse osmosis (RO)	Reverse osmosis (RO)		

Table 1. Classification of membrane processes according to their driving forces

Conventional membrane separation processes have at least two phase interfaces: feed fluid-membrane interface and product/permeate fluid-membrane interface on the two sides of the membrane. For example, commercialized membrane separation processes, such as RO, NF, UF, MF, GS, and PV, have two such phase interfaces. Over the last couple of decades, new membranes and membrane-separation techniques have appeared wherein the interface between two bulk phases is allowing the development of a new (improved) membrane, the creation of a novel membrane separation technique, the enhancement of the separation in existing membrane-separation processes, or the enhancement of the separation as such. The impact of such new techniques on conventional equilibrium-based separation processes/techniques is striking. The nature of the phase interface in such techniques is often crucial.

These innovative membrane systems, the separation principle of which is the phase equilibrium and known as membrane contactors, have been studied, realised, and used in integrated membrane processes. In Table 2, a classification of the different types of membrane contactors is given.

	Membrane distillation (MD)	Membrane strippers/ scrubbers	Osmotic distillation (OD)	Membrane emulsifiers	Membrane crystallizers (MCr)	Phase transfer catalysis
Phase 1	LIQUID	GAS/LIQUID	LIQUID	LIQUID	LIQUID	LIQUID
Phase 2	LIQUID	LIQUID	LIQUID	LIQUID	LIQUID	LIQUID
Driving force	PARTIAL PRESSURE GRADIENT	CONCENTR. GRADIENT	PARTIAL PRESSURE GRADIENT	PRESSURE GRADIENT	PARTIAL PRESSURE GRADIENT	CONCENTR. GRADIENT
Limit to transport	TEMPERAT. POLARIZ.	RESISTANCE IN MEMBR. Or LIQUID	CONCENTR. POLARIZ.	RESISTANCE IN MEMBR. Or LIQUID	TEMPERAT. CONCENTR. POLARIZ.	RESISTAN. IN MEMBR. OR LIQUID

Table 2. Classification of membrane contactors

At present, redesigning important industrial production cycles by combining various membrane operations suitable for separation and conversion units, thus realizing highly integrated membrane processes, is an attractive opportunity because of the synergic effects that can be attained.

In various fields, membrane operations are already the dominant technology. Their utilizations as hybrid systems, in combination with other conventional techniques or integrated with different membrane operations, is considered the way forward rationale applications.

In this context, interesting examples are in seawater desalination ; in wastewater treatment and reuse ; and in gas separation.

2. Seawater desalination

Sea and brackish water desalination has been at the origin of the interest for membrane operations, and the research efforts on RO membranes have had an impact on all of the progress in membrane science. Reverse osmosis desalination plants are currently leading the the desalination market, with RO installations representing 60% of the total number of worldwide plants, whereas thermal processes represent just 34.8% (Drioli et al, 2011). In Fig. 3 an example of the water desalination processes developed by the Japanese Water Re-Use Promotion Center, in co-operation with Takenaka Corporation and Organo Corporation: this process uses solar energy allowing the installation at location with no external electric energy supply (Drioli et al, 2011).

In Table 3 a list of traditional membrane technologies for water treatment is given.

The great flexibility, operational simplicity and mutual compatibility for integration of membrane operations offer the possibility of combining different membrane technologies for minimizing the limits of the single membrane units and for increasing the efficiency of the overall system.

Nowadays the most part of conventional seawater desalination plants use either RO or Multi-Stage Flash (MSF) technology. Thermal desalination is the most frequently applied technology in the Middle East, whilst membrane processes have rapidly developed and now surpass thermal processes in new plant installations due to the lesser energy consumptions (2.2.-6.7 kWh/m^3 for seawater RO, with respect to 17-18 kWh/m^3 for MSF).

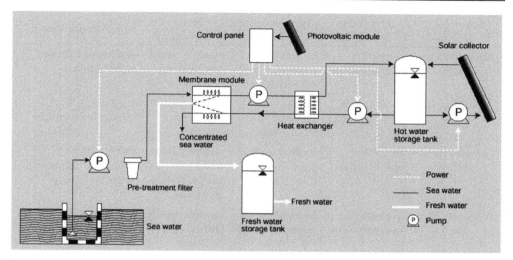

Fig. 3. Water desalination plant based on membrane process using solar energy, developed by the Japanese Water Re-use promotion center, in co-operation with Takenaka Corporation and Organo Corporation (Drioli *et al*, 2011).

However, high quality feedwater is required to ensure stable, long term performance, and an effective pre-treatment is essential for an efficient plant operation. In the past conventional, (i.e. conventional and physical) pre-treatment without the use of membrane technologies has been applied. Nowadays, membrane-based pre-treatments (such as MF, UF, NF) tend to replace conventional pre-treatment systems.

MF is a low energy-consuming technique extensively used to remove suspended solids and to lower chemical oxygen demand (COD)/biochemical oxygen demand (BOD) and silt density index (SDI). UF retains suspended solids, bacteria, macromolecules and colloids and despite of the larger pressure gradient with respect to MF, this membrane separation method is competitive against conventional pre-treatments. In an integrated membrane pre-treatment (Figure 4), the benefits of lower fouling rates of RO membranes compensates the higher cost membrane pre-treatment equipment.

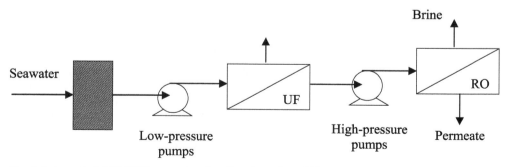

Fig. 4. Integrated (UF/RO) seawater desalination plant: UF as a pre-treatment

Separation process	Membrane type used	Applied driving force	Mode of separation	Applications
Microfiltration (MF)	symmetric macroporous	hydrostatic pressure 0.05-0.2 MPa	size exclusion, convection	water purification, sterilization
Ultrafiltration (UF)	Asymmetric macroporous	hydrostatic pressure 0.1-0.5 MPa	size exclusion, convection	separation of molecular mixtures
Diafiltration (DF)	asymmetric macroporous	hydrostatic pressure 0.1-0.5 MPa	size exclusion and dialysation, diffusion	purification of molecular mixtures, artificial kidney
Nanofiltration (NF)	asymmetric mesoporous	hydrostatic pressure 0.3-3 MPa	size exclusion, diffusion, Donnan-exclusion	separation of molecular mixtures and ions
Reverse osmosis (RO)	asymmetric skin-type, dense or microporous	hydrostatic pressure 1-10 MPa	solution-diffusion mechanism	sea & brackish ,water desalination
Electrodialysis (ED)	symmetric ion-exchange membrane	electrical potential	Donnan-exclusion	water desalination
Donnan Dialysis (DD)	symmetric ion-exchange membrane	concentration of ions	Donnan-exclusion	water softening
Membrane Distillation (MD)	symmetric porous hydrophobic membrane	Vapor pressure	diffusion	water desalination, concentration of solutions

Table 3. Membrane operations used for water treatments

Macedonio and Drioli (2010) analysed an integrated membrane-based desalination plant with membrane crystallization (MCr) as post-treatment for the recovery of salts and water contained in the NF/RO retentate streams of a desalination plant (Fig.5).

The feed water enters into MF membrane modules to be cleaned from suspended solids and large bacteria. After MF, process water is pressurized and then sent to NF membrane modules to be cleaned from turbidity, microorganisms, hardness, multivalent ions and 10-50% of monovalent species. After NF step, the RO step requires that the process water be pressurized to overcome the osmotic pressure. In the RO operation the process water is separated into a permeate and a brine. RO brine enters into the precipitator in which is mixed with Na_2CO_3 for the removal of the Ca^{2+} ions of the RO brine as $CaCO_3$. The process stream enters the MCr modules where it is separated into a permeate, a purge and a salt containing stream.

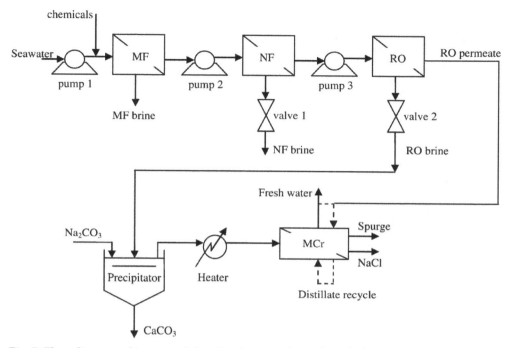

Fig. 5. Flow diagram of integrated desalination membrane-based plant (Macedonio and Drioli, 2010).

Hybrid membrane systems, in combination with conventional separation process, are advantageous in many industrial processes and in particular in desalination systems.

In recent years, the concept of simple hybrid multistage flash-reverse osmosis (MSF/RO) configuration has been applied to a number of existing or new commercial desalination plants. The SWCC Jeddah, Al-Jubail and Yanbu existing Power/Water cogeneration plants are expanded for more water production by combining with new SWRO desalination plants. The simple hybrid desalination arrangement enabled the increase of the water-to-power ratio and utilized effectively the available intake/outfall facilities.

The MSF and RO operate in parallel and are entirely independent. The water product of the single pass RO unit is blended with the MSF product (Hamed, 2006)

The 100 MGD desalination plant in Fujairah (UAE) is one of the largest hybrid MSF (Multistage Flash evaporators) /RO installation in the worlds: it combines a 62.5 MGD MSF and 37.5 MGD SWRO (Fig.6).

This hybrid desalination system is designed to provide significant operational savings by reducing fuel consumption by up to 25 per cent compared with a similar-sized plant based only on MSF technology. Other key criteria influencing the design of the desalination plant were feed water quality, product water requirements and compatibility with the cogeneration of electricity.

Fig. 6. Fujairah desalination plant in United Arab Emirates, from
http://fujairahinfocus.blogspot.com/2011/10/fujairah-power-and-desalination-plant.html

The water production system at the Fujairah desalination plant is comprised of five Doosan MSF units producing 57 million l/day (12.5 MGD) each and one RO unit with a design capacity of 171 million l/day (37.5 MGD). The RO unit was supplied by Ondeo Degremont. For drinking water supply, distillate from the MSF units and desalinated water from the RO plant are mixed in a distillate header and treated in a re-mineralization unit before passing into the potable water storage tanks. Prior to export to the water transmission line, potable water is stored in five potable water tanks, each with a capacity of 91 million l (20 million gallons).

A promising approach for pre-treatment of seawater make-up feed to MSF and SWRO desalination processes using NF membranes has been introduced by the R&D Center (RDC) of SWCC.

NF membranes are capable to reduce significantly scale forming ions from seawater, allow high temperature operation of thermal desalination processes, and subsequently increase water productivity.

The developed fully integrated systems NF/MSF and NF/SWRO/MSF result in high water productivity and enhance thermal performance compared to the currently used simple hybrid desalination arrangements (Hamed, 2006).

3. Wastewater treatment and reuse

Considerable advances in MF, UF and NF technologies to recover municipal wastewater have been also achieved. Also in this case, the implementation of integrated membrane systems is growing rapidly with excellent results.

In fig. 7 the main treatment steps to recover municipal wastewater from Kuwait City and the surrounding area are reported: A conventional biological wastewater treatment plant (WWTP) treats the effluent to better than secondary effluent quality. The secondary effluent then flows to the water reclamation plant, which uses UF and RO to further treat the water for reuse. Sludge from the wastewater treatment plant is treated to allow for disposal by

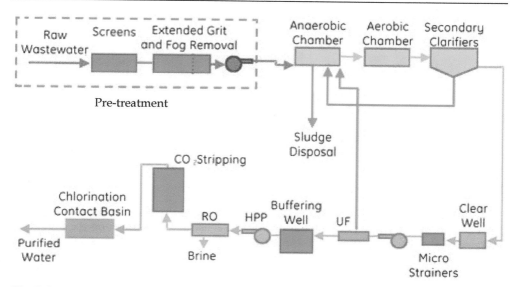

Fig. 7. Wastewater treatment facility at Sulaibiya near Kuwait City from http://www.water-technology.net/projects/sulaibiya/

landfill, incineration, or by composting. Membrane filtration was selected to provide robust pretreatment of the secondary-treated municipal effluent before being fed to the RO. Membrane filtration was chosen over conventional tertiary clarification and filtration because it reduced the plant chemical consumption and could guarantee that low turbidity water is fed to the RO. It is expected that better quality pre-treatment to the RO will lead to longer membrane life, lower operating pressure, and reduced cleaning frequency for the RO system. The UF plant utilizes Norit's X-Flow membranes, which are capillary hydrophilic hollow fibers. The UF units are operated individually. Each unit is backwashed regularly, whereby all suspended matter that is being retained by the membranes is removed from the plant. The backwash water is pumped back upstream of the WWTP to achieve the highest possible overall water recovery for the plant. The salinity of the municipal effluent has an average monthly value of 1,280 mg/l TDS, with a maximum value of 3,014 mg/l. RO is used to desalinate the water to 100 mg/l TDS, as well as provide a second barrier to bacteria and viruses. RO technology is well proven for desalinating municipal effluent. The system consists of 42 identical skids in a 4:2:1 array. Approximately 21,000 membrane modules, provided by Toray of America, were required for this project. The RO product passes through a stripper to remove carbon dioxide to adjust pH with a minimum amount of caustic before distribution, and the product is then chlorinated before leaving the plant. RO brine is disposed of into the Persian Gulf.

Membrane bioreactors (MBR) are a combination of activated sludge treatment and membrane filtration for biomass retention. Low-pressure membrane filtration, either UF or MF, is used to separate effluent from activated sludge. The two main MBR configurations involve either submerged membranes (Fig. 8) or external circulation (side-stream configuration).

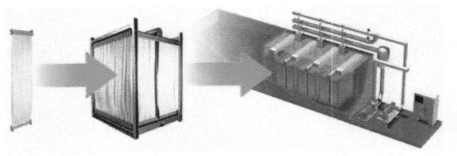

Fig. 8. Submerged membrane module for wastewater treatment. From ZeeWeed®
Submerged Membrane System, from http://www.gewater.com.

Since the early MBR installations in the 1990s, the number of MBR systems has grown
considerably. One key trend driving this growth is the use of MBR system for decentralized
treatment and water reuse. The successful introduction of MBR systems into small scale and
decentralized applications has led to the development of packaged treatment solutions from
the main technology suppliers. The company Conder Products, UK, designed the package
treatment plant Clereflo MBR; Zenon Environmental Inc., now a part of General Electric,
produced ZeeMod®.

The Pleiade® Plate & Frame membranes produced by Orelis©, France, which is one of
Europe's leading membrane manufacturer, are installed in skids mounted outside the
bioreactor and the sludge is circulated through the module in high speed, by pumps (Fig. 9).
This configuration and the Orelis membranes have several advantages over other MBR
systems.

Fig. 9. Pleiade® MBR membrane bioreactor for treating 1000 m³/day of effluent from
www.vic-ws.it/site/down/PLEENG0411.pdf

The membranes are in the UF range and offer high separation capabilities (0.02 μm,
MWCO=40 kD), unlike most MBR membranes which fall under the MF range. As a result,

the permeate quality is superb, even in very high sludge concentrations (up to 2%) and in difficult applications. Besides TSS, the membranes effectively reject also bacteria and even viruses.

High flow rate and pressure feed of the reactor content (MLSS) into the Pleiade membrane module, enable high permeate fluxes (60-80 lit/m2*hr). As a result, the membrane area required for treatment of a given WW flow is much lower than in most other MBR membranes.

High circulation speed over the membranes (2 m/sec) reduces fouling accumulation, and membrane cleaning (CIP) demand.

External membrane systems enable full modularity and easy expansion of WW treatment capacity. Gao et al (2011) developed a completely green process based on the integration of MBR with UF by treating micro-polluted source water in drinking water treatment. The removal of organic matter is carried by both a biodegradation mechanism in the MBR and by the MF/UF membrane, while the nitrification in MBR removes ammonia (Fig. 10).

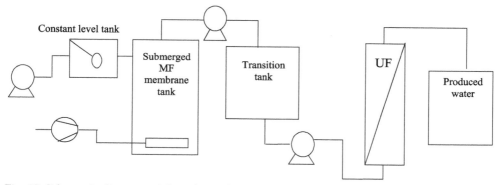

Fig. 10. Schematic diagram of the pilot scale experimental set-up MBR/UF (Gao *et al*, 2011)

The pulp and paper industry is one of the most water-dependent industries (Nurdan and Emre, 2010). The alkaline peroxide mechanical pulping (APMP) process has been widely applied especially in Asia (Liu *et al*, 2011) for the high yield, and the relatively low pollution. To achieve a closed wastewater loop, several APMP plants in the world have attempted to concentrate the total effluent by using a multi-effect evaporation system. Zhang *et al* (2011) studied a hybrid process UF/Multi-effect-evaporation (MEE) to concentrate effluent from APMP plants. With this new membrane concentration process, 88% of the water in the effluent can be removed, 1.4 bilion KWh power could be saved; the capital investment for MEE could also be decreased by 88% correspondingly.

The possibility of redesigning overall industrial production by the integration of various already developed membrane operations is of particular interest: low environmental impacts, low energy consumption, higher quality of final products and new available products are the advantages obtained.

The leather industry might be an interesting case study because of (i) the large environmental problems related to is operation; (ii) the low technological content of its traditional operations; (iii) the tendency to concentrate a large number of small-medium industries in specific districts. In Fig. 11 an ideal process based on integrated membrane operations for the tanning process is showed.

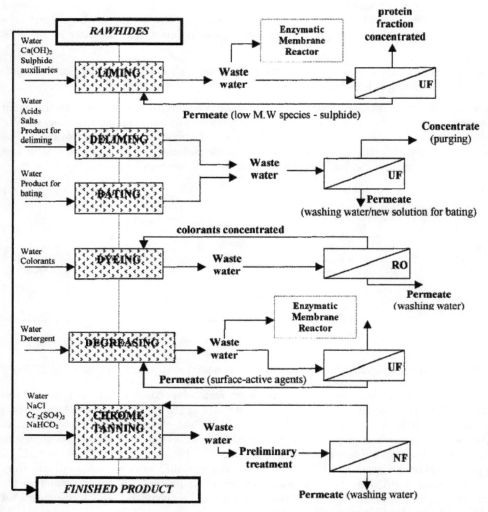

Fig. 11. Integrated membrane process proposed for the tanning process in the leather industry (Drioli & Romano, 2001)

The pollution problems of the leather industry have been minimized one by one at the point where they originate, thereby avoiding the need for huge wastewater treatment plants at the end of the overall production line. In addition, the membrane operations act by physical mechanisms without modification of the chemical procedure at the origin of the final high quality of the leather (Cassano *et al*, 2003)

4. Gas separation

In Table 4 the membrane sales involved for industrial gas separations are shown.

Application	Annual membrane system sales (10^6 U.S. \$)	
	Year 2010	Year 2020
Nitrogen from air	100	125
Oxygen from air	10	30
Hydrogen H_2/gas (CO, N_2, C_1, C_2) H_2/gas (C_{3+}, CO_2)	40 20	80 70
Natural gas CO_2 removal NGL removal and recovery N_2 removal, dehydration	60 20 10	100 50 25
Vapour (C_{2+})/Gas (N_2, Ar)	30	90
Vapour/Vapour (including dehydration)	20	100
Air dehydration/other	30	60
TOTAL	340	730
Annual growth, %	8	8

Table 4. Sales estimates and sales predicted for the principal gas and vapour separation applications

Nowadays, obtaining cheap high purity gases or enriched gas mixtures (the air, in particular) is a very important problem in industry and medicine as well as in everyday life (Bodzek, 2000).

Methods for air separation or oxygen enrichment can be divided into two groups (Freeman *et al*, 2006): cryogenic and non - cryogenic. The gaseous oxygen and nitrogen market is dominated by cryogenic distillation of air, and vacuum swing adsorption (Koros *et al*, 2000). Of the non – cryogenic methods, selective adsorption on zeolites and carbon adsorbents are available (Lin & Guthrie, 2006), and more and more attention is attracted to the membrane separation techniques (Dhingra & Marand, 1998) . The success of polymeric membranes has been largely based on their mechanical and thermal stability, along with good gas separation properties. The process of membrane separation is continuous, has a low capital cost, low power consumption, and the membranes, at least in gas separation, do not require regeneration (Vansant & Dewolfs, 1990).

Nitrogen production today is the largest GS process in use. Nitrogen gas is used in many applications (e.g., to prevent fires and explosions in tanks and piping systems and to prevent equipment degradation, during shutdown periods, in compressors, pipelines and reactors). Single-stage membrane operation is preferred. Air is pressurized and fed into the membrane separators; faster gases (O_2, CO_2, water vapor) permeate through the polymeric fiber walls, are collected and vented to the atmosphere while the slower, non-permeate N_2 gas is available at the other end of the separator.

Oxygen production by membrane systems is still underdeveloped, since most of the industrial O_2 applications require purity higher than 90%, which is easily achieved by adsorption or cryogenic technologies but not by single-stage membranes.

New materials are being developed that could possibly have higher permeabilities than conventional solid electrolytes. Promising oxygen permeation have been obtained in many perovskite systems (Zhu et al, 2009; Stiegel, 1999). The dense perovskite type membranes transport oxygen as lattice ions at elevated temperatures with infinite selectivity ratios in O_2 separations.

The oxygen-ion conducting membranes must operate above 700°C: an efficient and cost-effective way to recover the energy contained in the non- permeate, oxygen-depleted stream is that of integrating the membrane system with a gas turbine (Fig. 12) (Drioli & Romano, 2001).

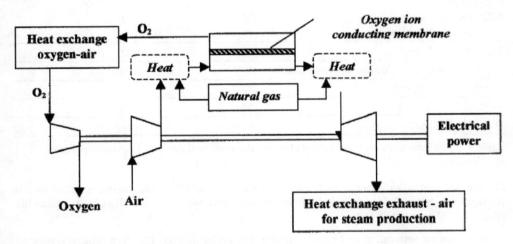

Fig. 12. Integrated system for O_2 and power productions (Drioli & Romano, 2001).

Hydrogen recovery was among the first large–scale commercial applications of membrane GS technology. The commercial success in the mid-1970's in Louisiana of the Permea hollow-fiber Prism system for in-process recycling of hydrogen from ammonia purge gases (Fig. 13) was the starting point of the penetration of membrane technology in large-scale manufacturing. A two-step membrane design was chosen for this ideal application for membrane technology: the ammonia reactor operates at high pressures (ca. 130 bar), thus providing the necessary driving force for separation; the H_2/N_2 membrane selectivity is high and the feed gas is free of contaminants.

This technology has been extended to other situations. In Table 5, some recent H_2 membrane applications are listed. H_2 recovery from refinery streams is an emerging field for membrane GS in the petrochemical industry; it is a key approach to meet the increased demand of hydrogen (for hydrotreating, hydrocracking or hydrodesulfurization processes) owing to new environmental regulations. An example is the H_2 recovery from high pressure purge gas of a hydrotreater.

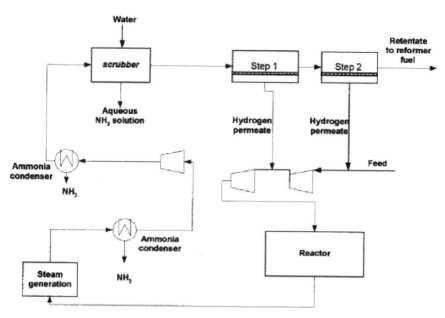

Fig. 13. Plant for H_2 recovery from ammonia synthesis (Drioli & Romano, 2001)

Refining	Chemical/Petrochemical
Catalytic reformer off gases	Synthesis gas composition adjustment (IGCC)
HDT/Recycle purge gases	Methanol purge gases
Refinery fuel gases	Ammonia purge gases
PSA off gases	Polypropylene purge
	Polyethylene purge
	Styrene off gases
	Coke oven gases
	Electrolysis gases
	Cyclohexane plant gases

Table 5. H_2 membrane applications

In a comparison of three separation technologies (membrane, PSA, cryogenic distillation) applied for H_2 recovery from refinery off-gas, Spillman (1989) reported that the use of membranes represent the best choice. An evaluation of these processes is provided in terms of sustainability indexes. The *Energy Intensity* of the membrane system is the lowest; the *Mass Intensity* is less than 50% of that for the conventional separations; the membrane system occupies a tenfold lower area required by PSA and cryogenic distillation of corresponding capacity (highest *Productivity/Footprint ratio*).

Carbon dioxide removal from natural gas (natural gas sweetening) is mandatory to meet pipeline specifications (e.g., down to 2% vol. in U.S.A.), since CO_2 reduces the heating value of natural gas, is corrosive and freezes at a relatively high temperature, forming blocks of dry ice that can clog equipment lines and damage pumps. Membrane technology is attractive for CO_2 and H_2S removal, because many membrane materials are very permeable to these species (enabling a high recovery of the acid gases without significant loss of pressure in the methane pipeline product gases), and because treatment can be accomplished using the high wellhead gas pressure as the driving force for the separation. A high natural gas recovery (>95%) can be achieved in multi-stage systems.

Cynara-NATCO produces hollow fiber modules for CO_2 removal and has provided a membrane system (16 in. modules) for the natural gas sweetening in an offshore platform in the Thailand gulf (830,000 Nm^3/h), which is the biggest membrane system for CO_2 removal. The Natco Group has been awarded in 2008 a \$24.9 million contract to provide membrane separation technology and equipment to capture CO_2 for re-injection in Bouri. Eni Oil Limited Libyan Branch operates the Bouri production platforms in the Mediterranean Sea, one of which will house the pre-treatment equipment and membrane systems intended to treat more than 160MM SCFD gas.

The Cynara(R) membrane system has been designed for the selective capture of CO_2 from the gas stream for this retrofit project on the existing platform. Natco will design, engineer, and fabricate four membrane skids and related valving, as well as pretreatment equipment. The contract was awarded in May 2008.

The integration of membranes with other well-established separation processes in the chemical and petrochemical industries, , was considered in different works (Doshi, 1987; Choe et al, 1987; Dosh & Dolan, 1995; Bhide et al, 1998). Usually, the combination membranes / PSA is considered in H_2 separation, while hybrid membranes + amine absorption are applied to the CO_2 separation.

A comparison of the separation cost for the membrane process with diethanolamine (DEA) absorption showed that the membrane process is more economical for CO_2 concentrations in the feed in the range 5–40 mol% (Bhide et al, 1998).

If membrane processes are not economically competitive because of the high H_2S concentration in the feed, the separation cost could be significantly lowered by using hybrid membrane processes. In the block diagram of Fig.14 the membrane unit removes two-thirds of the CO_2 and the amine plant removes the remaining. The combined plant is significantly less expensive than an all-amine or all-membrane plant.

A hybrid system (Cynara membranes + amine absorption) is operating since 1994 in Mallet (Texas, U.S.A.) to perform the bulk removal of CO_2 from associated gas (90% CO_2 and heavy hydrocarbons), before downstream treating. The membrane system offered a 30% reduction in operating cost when compared with a methyl diethanolamine (MDEA) system and significantly reduced the size of the subsequent operations (Blizzard et al, 2005).

An integrated system combining membrane permeation and pressure swing adsorption (PSA) has been developed for CO_2/N_2 gas separation (Esteves and Mota, 2007). By using the membrane as a pre-bulk separation unit and coupling it to the PSA, the separation performance of the hybrid scheme is enhanced with respect to that of the two stand-alone

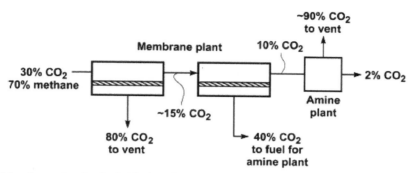

Fig. 14. Membrane/amine hybrid plant for the treatment of natural gas (Baker & Lokhandwala, 2008).

units. Instead of constant-composition regular feed, the PSA is fed with a mixture which is progressively enriched in the more adsorbed component during the pressurization and high-pressure adsorption steps of the cycle. This results in sharper concentration fronts. The hybrid has been applied successfully to the bulk separation of an 30:70 mol% CO_2/N_2 mixture over activated carbon. Process performance is reported in terms of product recovery and purity at cyclic steady state. Numerical simulations were validated by experimental work on a composite membrane and a laboratory-scale PSA unit.

MTR (Membrane Technology and Research, Inc.) has proposed an innovative retro-fit process for the post-combustion carbon capture and sequestration (CCS) of existing coal-fired power plants (Merkel et al., 2010). According to preliminary calculations, a two stage membrane process (Figure 15) should be able to recover 90% of the CO_2 released by the power plant at a competitive cost of 23$/ton CO_2 sequestrated. The selected lower cost configuration uses a blower upstream of the first membrane unit and a vacuum pump downstream in order to boost the driving force for the separation. At the same time, in the second membrane module the air to be fed to the boiler strips part of the CO_2 in the N_2 enriched flue gas from the first membrane module, thereby recycling CO_2 and building a higher CO_2 concentration for the first membrane module: as a consequence, the sequestered CO_2 flux in the first membrane unit is enhanced.

MTR is now involved in a contract with the US Department of Energy for the construction of a membrane skid containing MTR's Polaris membranes capable of 90% CO_2 capture from a 20 tons-of-CO_2/day slipstream of coal fired flue gas. The skid will be operated during a 6 month field test at Arizona Public Service's Cholla power plant. (http://www.mtrinc.com/news.html#DOE14mil).

In the process industry the final choice of a separation process is the result of a balance between the economics, the desired purity and recovery of the product and other conditions and restrictions, such as the desired capacity and the composition of the feed and the possibility of integration with other processes. In this logic, the integration of commercial membrane separation units in the recovery of ethylene or propylene monomer from polyolefin resin degassing vent streams has been proposed by Baker et al (1998). A simplified flow scheme of the process that generates such streams is shown in Fig. 16(a).

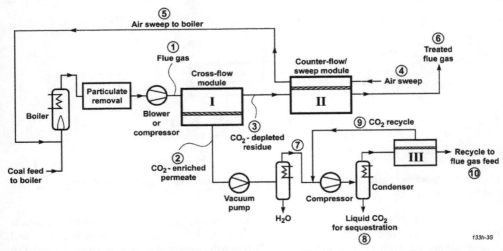

Fig. 15. Simplified flow diagram of a two-step counter-flow/sweep membrane process to capture and sequester CO_2 in flue gas from a coal-fired power plant. The base-case membrane with a CO_2 permeance of 1000 gpu and a CO_2/N_2 selectivity of 50 was used in the calculations (Merkel et al, 2010).

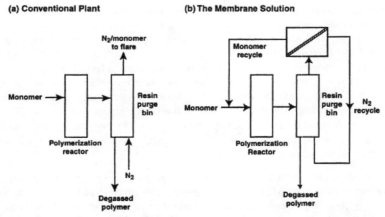

Fig. 16. A schematic diagram of monomer recovery and nitrogen recycle from polyolefin plant resin degassing operations: (a) conventional plant (b) membrane vapor gas separation system (Baker et al, 1998)

In a typical polyolefin polymerization plant, after polymerization, the polymer is removed to a low-pressure chamber. amounts of sorbed monomers and processing solvents, which must be removed before the polymer can be used. Therefore, the raw polymer is passed to large resin degassing bins through which nitrogen is circulated.

In the past, no economical method of separating these gases was available, so the stream was used as low grade fuel and the monomer content lost. The value of these potentially recoverable monomers in a typical polymerization plant is very high. The placement of a membrane to recover and recycle the gases is shown schematically in Fig. 16(b).

Membrane vapor gas separation systems have been installed worldwide. The main membrane applications in the petroleum industry are vapor recovery in tank farms and hydrogen recovery in refineries and chemical plants. Other developed applications of the membrane technology in refineries include solvent recovery in lube oil manufacturing (Max-DeWax developed jointly by Grace Division and Exxon Mobil) and aromatics removal from gasoline.

5. Conclusion

The development of innovative processes that follow the process intensification strategy for a sustainable industrial growth is critical to the production by non-polluting, defect-free and safe industrial processes. Membrane operations show a higher efficiency than conventional separation and reaction unit operations. They offer new options for the razionalization of innovative production cycles. Membrane engineering plays a crucial role in water desalination, in municipal water reuse (by MBR), in petrochemicals and in the field of gaseous separations. There are also some interesting opportunities to integrate membrane operations into existing industrial processes to achieve the benefits of process intensification.

6. References

Baker, R.W., Wijmans, J.G. and Kaschemekat, J.H. (1998). The design of membrane Vapor gas separation systems. *J. Membr. Sci.* Vol. 151 pp.55-62 ISSN 0376-7388

Baker, R. & Lokhandwala, K. (2008). Natural Gas Processing with Membranes: An Overview. *Ind. Eng. Chem. Res.* Vol. 47 No 7 pp.2109-2121 ISSN 0888-5885.

Bhide, B.D.; Voskericyan, A.; Stem, S.A. (1998) Hybrid processes for the removal of acid gases from natural gas. *J. Membr. Sci.* Vol. *140* No 1 pp.27-49 ISSN 0376-7388

Blizzard, G.; Parro, G.; Hornback, K. (2005). Mallet gas processing facilities uses membranes to efficiently separate CO_2. *Oil & Gas Journal* Vol. 103 No 14 pp 48-53.

Bodzek, M. (2000) Membrane techniques in air cleaning. *Pol. J. Environm. Stud.* Vol. 9. No.1 , pp. 1-12 ISSN 1230-1485.

Cassano, A; Adzet, J; Molinari, R.; Buonomenna, M.G.; Roig, J; and Drioli, E. (2003) Membrane treatment by nanofiltration of exhausted vegetable tannin liquors from the leather industry. *Water Research.* Vol. 37, pp. 2426-2434 ISSN 0043-1354.

Charpentier, J.C. (2007). In the frame of globalization and sustainability, process intensification, a path to the future of chemical and process engineering (molecules into money). *Chemical Engineering and Processing: Process Intensification,* Vol. 134, No. 1-3, pp.84 92, ISSN 0255-2701

Choe, J.S.; Auvil, S.R.; Agrawal, R. (1987) Process for separating components of a gas stream" US Patent 4,701,187

Cobb, C.; Schuster, D.; Beloff, B.; Tanzil, D. (2007). Benchmarking Sustainability.Chemical Engineering Progress, Vol. 104, No.6, pp. 38-42, ISSN 0360-7275

Dhingra , S.S. & Marand, E. (1998) Mixed gas transport study through polymeric membranes , *J. Membr.Sci.* Vol. 141 No 1 pp 45-63 ISSN 0376-7388

Doshi, K.J. (1987) Enhanced gas separation process, US Patent 4,690,695, Union Carbide.

Doshi, K.J.; Dolan, W.B. (1995) Process for the rejection of CO_2 from natural gas", US Patent 5,411,72, UOP.

Drioli, E; Romano, M. (2001). Progress and New Perspectives on Integrated Membrane Operations for Sustainable Industrial Growth. *Ind. Eng. Chem. Res.* Vol. 40, pp. 1277-1300, ISSN 0888-5885

Drioli, E; Stankiewicz, A; Macedonio, F. (2011). Membrane Engineering in process intensification-An overview. *J. Membr. Sci.* Vol. 380, pp. 1-8, ISSN 0376-7388

Esteves, I.A.A.C., Mota, J.P.B. (2007). Hybrid Membrane/PSA Processes for CO_2/N_2 Separation. *Adsorption Science & Technology* Vol. 25, pp. 693-715 ISSN 0263-6174.

Freeman, B; Yampolskii, Y; Pinnau, I. (2006) Eds., Materials Science of Membranes for Gas and Vapor Separation, John Wiley and Sons, Ltd., Chichester ISBN 9780470029039.

Gao, W; Liang, H; Wang, L.; Chang, H.-q, , Li, G-b. (2011) Pilot Study of Integrated MF based MBR and UF for Drinking Water Production by Treating Micropolluted Source Water. *IPCBEE* Vol. 14, pp. 11-16, ISSN 2010-4618

Hamed, O.A. (2006) Overview Of Hybrid Desalination Systems - Current Status And Future Prospects. *Desalination* Vol. 186, pp. 207–214, ISSN 0011-9164

Koros, W.J.; Mahajan, R.(2000) Pushing the limits on possibilities for large scale gas separation: which strategies? *J. Membr. Sci.* Vol. 175 No. 1 pp. 181-196 ISSN 0376-7388

Lin, L. & Guthrie, J.T. (2006) Novel oxygen - enhanced membrane assemblies for biosensors. *J. Membr. Sci.* Vol. 278 No 1-2 pp 173-180. ISSN 0376-7388

Macedonio, F.; Drioli, E. (2010). An exergetic analysis of a membrane desalination system. *Desalination* Vol. 261, pp. 293-299, ISSN 0011-9164

Merkel, T.C.; Lin, H.; Wei, X.; Baker, R. (2010). Power plant post-combustion carbon dioxide capture: an opportunity for membranes, *J. Membr. Sci.* Vol. 359 No 1-2 pp 126-139. ISSN 0376-7388

Nurdan, B.; Emre, K. (2010) Economic evaluation of alternative wastewater treatment plant options for pulp and paper industry. *Sci. Total. Environ.* Vol. 408, No.24, pp.6070-6078, ISSN 0048-9697.

Spillman, R. (1989) Economics of gas separation by membranes. *Chem.Eng. Prog.* Vol. 85, pp.41-62, ISSN 0360-7275.

Stiegel, G. J. (1999) Mixed conducting ceramic membranes for gas separation and reaction. *Membr. Technol.* Vol. 1999 No 110 pp 5-7 ISSN 0958-2118

United Nations General Assembly (1987). Report of the World Commission on Environment and Development: Our Common Future

Vansant, E.F. & Dewolfs, R. (1990) *Gas Separation Technology*, Elsevier, 1990 ISBN 0444882308

Zhang, Y; Cao, C-Y; Feng, W-Y; Xue, G-X; Xu, M. (2011) Performance of a pilot scale membrane process for the concentration of effluent from alkaline peroxide mechanical pulping plants. *BioResources*. Vol.6, No.3, pp. 3044-3054, ISSN 19302126.

Zhu, X; Sun, S; Cong, Y; Yang, W. (2009) Operation of perovskite membrane under vacuum and elevated pressures for high-purity oxygen production. *J. Membr. Sci.* Vol.345 pp 47-52 ISSN 0376-7388.

10

Thermal Study on Phase Transitions of Block Copolymers by Mesoscopic Simulation

César Soto-Figueroa[1], Luis Vicente[2]
and María del Rosario Rodríguez-Hidalgo[1,*]
[1]*Departamento de Ciencias Químicas,*
Facultad de Estudios Superiores Cuautitlán,
Universidad Nacional Autónoma de México (UNAM)
[2]*Departamento de Física y Química Teórica,*
Facultad de Química Universidad Nacional Autónoma de México (UNAM)
México

1. Introduction

The block copolymers are an exceptional kind of macromolecules constituted by two or more blocks of different homopolymer chains linked by covalent bonds. These polymeric materials have received much attention over past few years due in large part to their ability to self-assemble in the melted state or in a selective solvent inside a variety of ordered phases or well-defined structures of high regularity in size and shape with characteristic dimensions between 100 and 500 nanometres. These ordered phases and their structural modification are the key to many valuable physical properties which make block copolymers of great industrial and technological interest. The molecular self-assemble and formation of periodic phases in the block copolymers depend of the strength of interblock repulsion and composition, for example, mesoscopic studies of the poly(styrene)-poly(isoprene) (PS-PI) diblock copolymer, have demonstrated that this synthetic material, may generate a series of long-range ordered microdomains when exist a weak repulsion between the unlike monomers isoprene and styrene, as result, the PS-PI diblock copolymer chains tend to segregate below some critical temperature, but, as they are linked by covalent bonds, the phase separation on a macroscopic level is prevented, only a local microphase segregation occurs (Soto-Figueroa et al., 2005, Soto-Figueroa et al., 2007). The phase transition from homogeneous state of polymeric chains to an ordered state with periodic phases is called microphase separation transition (MST) or order-disorder phase transition (ODT) (Leibler, 1980). The phase segregation and generation of ordered structures in the microscopic level of a diblock copolymer via an order-disorder phase transition is illustrated in Fig. 1.

As result of microphase segregation process, the block copolymers can display ordered structures constituted by homopolymer domains that haves only mesoscopic dimensions corresponding to the size of singles blocks. The microphase separation leads to different classes of well-defined periodic structures in dependence on the ratio between the degrees of

---

* Corresponding Authors

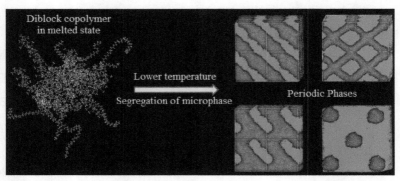

Fig. 1. The microphase segregation process occurs when the PS-PI diblock copolymers in melted state are transformed to a periodic inhomogeneous phase of ordered structures when the temperature diminishes.

polymerization of the component blocks. Periodic phases with specific morphologies such as: spherical, perforated layers, cylindrical, lamellar and Gyroid can be generated manipulating the composition or length of the component blocks (Strobl, 1997; Bates & Fredrickson., 1990).

The order-disorder and order-order phase transitions play an important role in the design and modification of new supramolecular materials and are the key to manipulate the physical and mechanicals properties in these polymeric materials.

In this chapter, attention has been concentrated on the order-disorder and order-order phase transitions that display the PS-PI diblock copolymers and in the mesoscopic simulations methods employed to explore the kinetics transformation pathway of well-defined ordered phases.

1.1 Order-disorder phase transition (ODT)

The order-disorder phase transition is a thermodynamic process controlled by enthalpic and entropic interactions, for example, when a diblock copolymer of type A-B are in a melted state by temperature effect, exhibits a homogeneous phase where all different block segments are completely miscible, in the reverse case when the different block segments are immiscible due to decrease of the temperature displays an heterogeneous state of ordered microphases: this conditions are describes by Gibbs free energy equation of mixing when $\Delta G_m < 0$ and $\Delta G_m > 0$ respectively.

$$\Delta G_m = G_{AB} - (G_A + G_B) \tag{1}$$

where G_A, G_B and G_{AB}, denote the Gibbs free energy of A and B segments in separate states and the mixed state, respectively. Equation (1) in accordance with Flory-Huggins theory can be expressed also as a sum of two thermodynamic contributions (Flory, 1953):

$$\Delta G_m = -T\Delta S_m + \Delta H_m \tag{2}$$

In this equation, ΔH_m and $T\Delta S_m$ exhibits the enthalpic and entropic interactions of mixing at temperature T. The entropic and enthalpic interactions of mixing of two component segments of diblock copolymer are given by:

$$\frac{\Delta S_m}{\overline{R}} = \overline{n}_A \ln \phi_A + \overline{n}_B \ln \phi_B \tag{3}$$

$$\Delta H_m = \overline{R}T \frac{V}{\overline{v}_c} \chi \phi_A \phi_B \tag{4}$$

The equations 3 and 4 can be rewritten as:

$$\Delta S_m = -K \left[n_A \ln \phi_A - n_B \ln \phi_B \right] \tag{5}$$

$$\Delta H_m = kT \chi N \phi_A \phi_B \tag{6}$$

where ϕ_A and ϕ_B are volume fraction of A and B components of diblock copolymer, $N=n_A+n_B$ denotes the total number of molecules or degree of polymerization and χ is the Flory-Huggins interaction parameter. The phase behaviour that exhibits the diblock copolymers during an order-disorder phase transition (microphase segregation) is controlled by both entropic and enthalpic interactions. The enthalpic interactions imply the repulsion magnitude between different species via Flory-Huggin's interaction parameter (χ), which represent the chemical incompatibility between different repetitive units and its magnitude is expressed by the type of monomers which integrate the diblock copolymer and has a strong dependence with the temperature:

$$\chi \approx \frac{1}{T} \tag{7}$$

the phase segregation behaviour is controlled by the value of χ, in this way, positives values of interaction parameter lead to incompatibility between different segments and the entropic interactions (ΔS_m) appears to be mostly positives, this generates a positive heat of mixing and therefore a $\Delta G_m > 0$. Negative values of χ lead to homogeneous state and therefore a $\Delta G_m < 0$.

Whereas the entropic interactions involve the configurational and translation displacement of polymeric chains, and are regulated through the degree of polymerization N, architecture constrains and blocks composition (Bates & Fredrickson., 1990, 1999; Hamley, 1998; Balta-Calleja et al., 2000; Thomas et al., 1995). The microphase segregation degree in the diblock copolymers depends of the enthalpic-entropic balance represented by the reduced parameter χN, the ODT occurs at a critical value of χN, the melt phase behaviour is thus governed by composition and a reduced parameter (Leibler et al., 1980; Bates et al., 1990). Three segregation regimes have been identified by Matsen and Bates and have been defined depending on the extent of microphase segregation: the weak ($\chi N \approx 10$), intermediate ($\chi N > 12\text{-}100$) and strong segregation regimes ($\chi N > 100$) (Matsen & Bates, 1996; Bates & Fredrickson, 1990). In the weak segregation regime, the volume fraction of one of the block varies sinusoidally about the average value generating the formation of ordered microphases, this regime is characterized by a diffuse interface between different components and is capable of to be modified by composition effect or for temperature effect (Hamley, 1998; Bates, 1991). In the intermediate segregation regime the composition profile becomes sharper generating ordered microphases with a narrow interface between blocks. The strong segregation regime due to saturation of the blocks composition contains essentially pure components; in this regime, the phase behaviour depends largely on the copolymer composition. Within these segregation regimes it is possible to predict and modify the phase behaviour of block copolymers given χ, N and the segment length (block composition).

A general description of phase behaviour that exhibits the well-defined ordered structures of the PS-PI diblock copolymer via order-disorder phase transitions have been explored by mesoscopic simulation methods. When PS-PI diblock copolymer is in melted state, the polymeric chains assume the lowest free energy configuration, if the diblock copolymer is cooled, the repulsion magnitude expressed by reduced parameter χN increases, when the value of this parameter exceeds a certain value specific ($\chi N \approx 10.5$) for the system under consideration, well-defined periodic structures evolves in the disordered state (Soto-Figueroa et al., 2005).

When the PS-PI diblock copolymer has a symmetric composition (volume fraction of both components are the same) display an ordered phase with lamellar morphology (LAM), Fig. 2(a), however, if the volume fraction of a component increases in relative to other component (asymmetric copolymer), the interface tends to become curve. In this case, the conformational entropy loss of the majority component is too high. Therefore, to gain the conformational entropy, the chains of the majority component tend to expand along the direction parallel to interface. As a result, the PS/PI interface becomes convex towards the minority component. This interface curvature effect is more pronounced when the composition of the diblock copolymer is more asymmetric, Fig. 2(b). The PS-PI diblock copolymers with asymmetric compositions can generate a wide range of ordered structures such as the body-centred-cubic (BCC), hexagonal packed cylinders (HPC), ordered bicontinuous double diamond (OBDD or Gyroid) and lamellar (LAM) arrangement via an order-disorder phase transition process, Fig. 2(c-f).

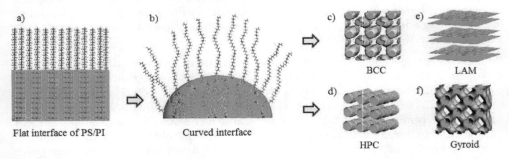

Fig. 2. Schematic representation of composition effect in the phase behaviour of the PS-PI diblock copolymer: a) lamellar phase (symmetric copolymer), b) curvature of PS/PI interface by effect composition (asymmetric copolymer), c) spherical phase, d) cylindrical phase, e) lamellar phase and f) OBDD phase.

The ordered phases with well-defined morphologies that shows the block copolymers are usually describes by the volume fraction of one block (f), the overall degree of polymerization (N), and Flory-Huggins interaction parameter (χ) (Soto- Figueroa et al., 2007). The periodic phases of type BCC, HPC, Gyroid and LAM explored by means of mesoscopic simulations are described in the following sections.

1.1.1 Body centred cubic phase

The body centred cubic (BCC) phase or spherical phase is a classic structure of great thermodynamic stability that exhibits the diblock copolymers of type A-B. This ordered structure shows two specific symmetries: a four-fold symmetry ([100] projection) and the hexagonal symmetry ([111] projection) of cubic array respectively, see Fig. 3.

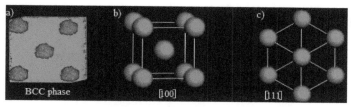

Fig. 3. Snapshots of BCC phase of PS-PI copolymer: a) ordered microdomains of PS and PI chains (red and green regions represent the PS and PI microdomain respectively), b-c) projections [100] and [111] of poly(styrene) microdomains (the poly(isoprene) matrix has been removed for a better visualization).

The BCC phase of diblock copolymers have been studied extensively by both theoretical and experimental methods. For example, the theoretical studies of PS-PI diblock copolymer with linear architecture have demonstrated that this periodic phase can be generated in a composition interval of 0.1 to 0.19 (volume fraction of poly(styrene)) with a stable four-fold symmetry (Soto-Figueroa et al., 2005). The spherical phase of this diblock copolymer is known as a stable phase, however, can be transformed on other ordered structure via an order-order phase transition.

1.1.2 Hexagonal packed cylinders phase

The hexagonally packed cylinders (HPC) phase is also considered as a classic phase of great thermodynamic stability due to its high packing. This ordered structure is characteristic of PS-PI diblock copolymers with linear architecture and can be generated via an order-disorder phase transition in a composition interval of 0.2 to 0.26 (volume fraction of poly(styrene), see Fig. 4.

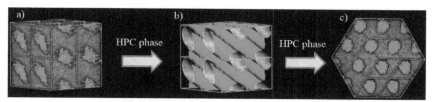

Fig. 4. Snapshots of HPC phase of PS-PI diblock copolymer: a) cylindrical microphase constituted by ordered microdomains of poly(styrene) and poly(isoprene), b) representation of density isosurfaces of microdomains of poly(styrene) and c) poly(styrene) cylinders arranged in a hexagonal lattice.

The HPC phase displays a tetrahedral arrangement of epitaxially cylinders arranged in a hexagonal lattice. In this ordered phase, the cylindrical microdomains of poly(styrene) exhibits an hexagonal arrangement of undulated microdomains immersed in a poly(isoprene) matrix.

1.1.3 Ordered bicontinuous double diamond phase

The ordered bicontinuous double diamond (OBDD) phase (or Gyroid phase) is the ordered structures more complex that exhibits some diblock copolymers of type A-B. For example, The PS-PI diblock copolymer can generate this ordered structure via an order-disorder phase transition, see Fig. 5.

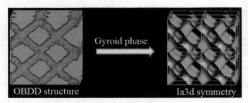

Fig. 5. Snapshots of OBDD arrangement of PS-PI diblock copolymer obtained by mesoscopic simulations: a) ordered microdomains of poly(styrene) and poly(isoprene) chains, and b) representation of density isosurfaces of microdomains of poly(styrene) (Soto-Figueroa et al., 2008).

The OBDD phase has a tetrahedral arrangement of epitaxially cylinders interconnected by channels of Ia3d symmetry (Hajduk et al., 1995). The OBDD phase of PS-PI copolymer exists only in a narrow interval of values (0.37 to 0.4 volume fraction of poly(styrene)) between the regimes of the perforated layer and lamellar phases. The Gyroid arrangement is known as a stable phase, however, can be transformed on other ordered structure via an order-order phase transition.

1.1.4 Lamellar phase

Ordered phase with lamellar morphology are characteristics of block copolymers of symmetric and asymmetric composition. This periodic phase is constituted of alternating layers of different homopolymer microdomains separates by flat interfaces, see Fig 6. For the case of PS-PI diblock copolymer with linear architecture, this displays the lamellar arrangement in a specific predominance interval between 0.47 to 0.68 volume fraction of poly(styrene).

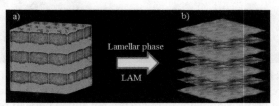

Fig. 6. Snapshots of lamellar phase of PS-PI diblock copolymer, the lamellar structure exhibits undulated microdomains of poly(styrene) and poly(isoprene): a) ordered microdomains of poly(styrene) and poly(isoprene) and b) representation of density isosurfaces of LAM phase.

The lamellar phase can be generated in the three segregation regimes ($\chi N \approx 10$, $\chi N > 12$-100 and $\chi N > 100$) via order-disorder phase transitions and can displays flat and undulated interfaces.

1.1.5 Phase diagram of PS-PI diblock copolymer

The ordered structures of PS-PI diblock copolymer represent under specific conditions the states with the lowest Gibbs free energy, these equilibrium phases can be classified through a phase diagram, Fig. 7. The diagram phase depicts the two regions associates constituted by a homogeneous state and an ordered state of five ordered phases different. Each equilibrium phase shows its predominance zone in terms of the volume fraction of poly(styrene) block and reduced parameter χN.

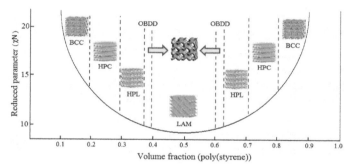

Fig. 7. Phase diagram of PS-PI diblock copolymer, the continuous curve describes the points of phase transition between the homogeneous state and the microphase-separated states. The ordered states split into different classes (BCC, HPC, G, HPL and LAM); the dashed lines show the predominance interval between the different types of ordered phases.

The segregation state in the phases diagram is controlled by the product χN, if χN is mayor than a critical value (typically of the order of 10.5), entropic factors dominate and the diblock copolymers exist in an ordered phases state. On the other hand, an order-to-disorder transition takes place for large values of χN. The phase behaviour of each ordered structures kind on the phase diagram of poly(styrene)-poly(isoprene) diblock copolymer can be modified through order-order phase transitions, this will be tackled in the following section.

2. Order-order phase transition in block copolymers

The diblock copolymers display a wide variety of classic phases with morphologies or defined structures such as: body centred cubic, hexagonal packed cylinders, hexagonal perforated layers, lamellar and the ordered bicontinuous double diamond phase, the morphology of these periodic structures can be controlled and modified by two different routes: 1) order-disorder phase transition and 2) thermotropic order-order phase transition.

In the order-disorder phase transition as was mentioned in the previous section, the phase behaviour in the block copolymers is governed by three experimentally controllable factors during the synthesis process: i) the overall degree of polymerization, ii) architecture of block copolymer, and iii) the interaction parameter between components blocks, of this thermodynamic process can emerge periodic structures with well-defined morphologies (Soto-Figueroa et al., 2005, Soto-Figueroa et al. 2007).

Other way to modify the phase behaviour in the block copolymers is through the temperature; thermally induced phase transitions have the potential to promote the kinetic control in these synthetic materials. Leibler was the first in predicting the phase transition between different ordered structures by temperature effect (Leibler et al., 1980). The thermally induced phase transformations are governed for anisotropic composition fluctuation effects (Ryu et al., 1999), when an ordered phase of specific morphology is subject to a thermal heating process, the homopolymer chains into the ordered microdomains exhibit thermodynamic instability by temperature effect and become less rigid, this entropic process promotes the polymeric chains movement (composition fluctuations) into the homopolymer microdomains and modifies with the time the shape of ordered phase.

Three segregation regimes have been defined to explain the extent of microphase segregation and the thermodynamic stability in the classical phases that exhibits the diblock copolymers, these segregation regimes are : weak regime ($\chi N \approx 10$), intermediate regime (χN >10-100) and strong segregation regime (χN >100) (section 1.1).

Our interest is concentrated in the weak segregation regime, because in this predominance zone, the diblock copolymers are characterized by a widened interface due to enhanced phase mixing. In the vicinity to this regime, thermotropic phase transition between different kinds of ordered phases can be generating (Bates et al., 1990; Matsen et al., 1996).

The transition from one ordered state to another is nowadays denominated as an order-order phase transition (OOT) (Sakurai et al., 1993; Kim et al., 1998; Almadal et al., 1992; Sakurai et al., 1996; Sakamato et al., 1997; Modi et al., 1999). In this way, the ordered phases that exhibit the diblock copolymers inside weak segregation regime can exhibit order-order phase transitions when are subjects to thermal heating cycles.

In the past decade the order-order phase transitions that display the classical phases of diblock copolymers (BCC, HPC, Gyroid and Lamellar) have been investigated by both experimental and theoretical studies. The order-order phase transition is thermoreversible process that develops transient metastable states during phase transformation. The order-order phase transition that exhibits the classical phases with specific morphology such as Gyroid, cylindrical, and lamellar are described next.

2.1 Order-order phase transition of HPC structure

The ordered phase of hexagonal packed cylinders also known as cylindrical structure is characteristic of diblock and triblock copolymers, for example, the poly(styrene)-poly(isoprene) diblock copolymer with linear architecture can generate this ordered arrangement in a specific composition of 0.2/0.8 (volume fraction) of PS/PI via an order-disorder phase transition. When the HPC phase of this diblock copolymer is subject to thermal heating cycles, exhibits an order-order phase transition to body-centred-cubic (BCC) structure (or spherical phase). The order-order transition pathway that exhibits the HPC to BCC phase was recently reported (Modi et al., 1999; Krishnamoorti et al., 2000).

Experimental and theoretical studies have confirmed that HPC phase develops transient metastable phases during thermal heating process (Kimishima et al., 2000; Krishnamoorti et al., 2000; Rodríguez-Hidalgo et al., 2009). The order-order transition pathway from HPC to spherical phase is sketched schematically in Fig. 8.

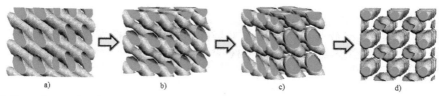

a) b) c) d)

Fig. 8. Snapshots of order-order phase transition process from HPC phase to spherical arrangement monitored by means of mesoscopic simulations: a) cylindrical phase, b) undulation process of poly(styrene) microdomains by temperature effects, c) breakdown of cylindrical microdomains of poly(styrene) and d) formation and stabilization of body-centred-cubic arrangement (BCC phase) (Soto-Figueroa et al., 2008).

Three transitory stages are typical of OOT process from HPC to cylindrical phase: (Stage 1) undulation process of cylindrical microdomains of poly(styrene) in the poly(isoprene) matrix. In this transitory stage the thermal heating induce the thermodynamic instability of HPC phase, where the cylindrical microdomains and the PS/PI interface becomes less rigid due to anisotropic composition fluctuations, this generates the undulation of the poly(styrene) microdomain in the poly(isoprene) matrix. (Stage 2) breakdown of cylindrical microdomains of poly(styrene) by temperature effects. When anisotropic composition fluctuations reach a critical point of thermodynamic instability the cylindrical microdomains become unstable and therefore they are broken in ellipsoids, (Stage 3) formation and stabilization of BCC phase. In this thermally induced stage, the ellipsoids microdomains generated by breakdown of cylindrical microdomains of poly(styrene) evolve to an equilibrium state where the uniform spherical microdomains are stabilized into body-centred-cubic arrangement. The order-order phase transition from HPC to spherical phase is a thermoreversible process. The inverse process from BCC state to HPC arrangement involves deformation and elongation of spheres into ellipsoids and coalescence of ellipsoids into the cylindrical microdomains. This process is driven by thermodynamic instability of the spherical interface caused by decrease of temperature.

2.2 Order-order phase transition of Gyroid structure

The Gyroid phase as was mentioned in the section 1.1.3 exhibits a tetrahedral arrangement of epitaxially cylinders interconnected by channels of Ia3d symmetry and is capable of suffering order-order phase transitions when it is submitted to thermal heating cycles. The poly(styrene)-poly(isoprene) diblock copolymer displays this ordered phase in a narrow interval of specific composition. The thermally-induced phase transition from Gyroid to lamellar phase in PS-PI diblock copolymer was explored in detail by mesoscopic simulation methods (Soto-Figueroa et al., 2008). The phase transformation during thermal annealed proceed in several stages via the generation two metastable phases (HPL and cylindrical). The order-order phase transition process from Gyroid to lamellar phase is sketched schematically in Fig. 9.

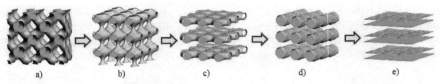

Fig. 9. Transitory stages and metastable phases generated during the order-order phase transition from Gyroid phase to lamellar arrangement: a) undulation of interconnected microdomains of poly(styrene) into the poly(isoprene) matrix, due thermal heating, b) breakdown of interconnected microdomains of Gyroid phase, c) formation of HPL metastable phase, d) formation of cylindrical metastable phase and e) lamellar phase.

When the Gyroid phase of PS-PI copolymer is subject to thermal heating cycles, the interconnected microdomains of poly(styrene) display an undulation process due to anisotropic composition fluctuations, Fig 9(a), with the increase of temperature the anisotropic composition fluctuations induce the breakdown of side interconnections in the Gyroid arrangement, Fig. 9(b), generating the first metastable phase with HPL arrangement, Fig. 9(c). The HPL metastable phase evolves later to cylindrical arrangement by temperature effect, in this transient stage the interconnections of perforated layer microdomains diminish their volume up to the breakdown of interconnections in the HPL arrangement and the formation of cylindrical phase (second metastable phase), Fig 9(d). Finally, the cylindrical metastable phase also changes over time due to anisotropic composition fluctuations. The cylindrical microdomains in this metastable phase are thermodynamically unstable, in order to reach a thermodynamic stability of minimal energy, the cylinders microdomains are joined together to evolve into undulating lamellar phase, Fig. 9(e), the lamellar phase consisting of alternating layers of PS and PI microdomains.

2.3 Order-order phase transition of lamellar structure

The lamellar phase is the ordered structure more simple that exhibit the multiblock copolymers is considered a classical phase of great thermodynamic stability. The lamellar phase consisting of alternating layers of different homopolymer microdomains. The PS-PS diblock copolymer exhibits this ordered arrangement in a specific composition interval (see section 1.1.4). When the lamellar phase of this diblock copolymer with an equivalent composition between their constituent blocks (0.5 volume fraction of poly(styrene) and poly(isoprene)) is subject to thermal heating cycles, it does not generate an order-order phase transition, the lamellar arrangement in this case evolve to a homogeneous state (melted), see Fig. 10.

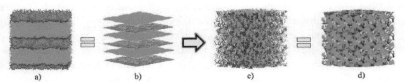

Fig. 10. Snapshots of thermal heating of lamellar phase (diblock copolymer of symmetric composition): a-b) equilibrium phase of lamellar structure (representation of diblock chains and surface isodensities of PS and PI microdomains) and c-d) homogeneous state after of thermal heating (melted phase).

Nevertheless, the lamellar phase generated of block copolymers of asymmetric composition, for example of 0.45/0.55 of PS/PI (volume fraction), it can exhibit a phase transformation to HPL arrangement when is subject to thermal heating via an order-order phase transition process (Mani et al., 2000). The order-order phase transition from lamellar to HPL phase is showed in Fig. 11.

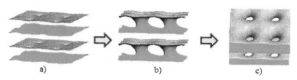

Fig. 11. Order-order phase transition stage from lamellar phase to HPL arrangement: a) undulation process of lamellar microdomains, b) interconnection of parallel microdomains of poly(styrene) and c) formation and stabilization of HPL arrangement.

The order-order phase transition process of this ordered arrangement is governed by the thermodynamic instability between the components microdomains, generated by thermal heating, where the anisotropic composition fluctuations play an important role in the phase transformation. In the initial stage the lamellar microdomains exhibit an undulation process due to thermal heating, Fig. 11(a), with the temperature increase, the lamellar phase becomes thermodynamically unstable, in order to reach a thermodynamic stability of minimal energy, the lamellar alternate microdomains of poly(styrene) are interconnected by means of narrow microdomains (parallels to PS/PI interface), finally the HPL phase evolve in an energetic equilibrium state (Soto-Figueroa et al., 2007), see Fig. 11(b-c).

3. Thermal study of double directionality of order-order phase transitions of hexagonally perforated layers (HPL) phases by mesoscopic simulation

During the past two decades have been reported theoretical and experimental studies of diblock copolymers, where the ODT and OOT transitions play a main role in the structure control and consequently in the physical properties control of these polymeric materials (Kimishima et al., 2000; Kim et al., 2006; Bodycomb et al., 2000; Krishanamoorti et al., 2000; Court et al., 2006; Soto-Figueroa et al., 2008). It is well-known that the poly(styrene)-poly(isoprene) diblock copolymer exhibits a wide variety of classical phases with specific morphologies such as: BCC, HPC, OBDD, LAM and HPL. The order-order phase transitions that exhibit this ordered structures with definite morphologies have been explored by Soto-Figueroa and Rodríguez-Hidalgo, they have confirmed the OOT´s between HPC to BCC, Gyroid to LAM microphase through mesoscopic simulations (Soto-Figueroa et al., 2007, 2008; Rodríguez-Hidalgo et al., 2009), although the OOT´s have been well investigated for the majority of classical phases, the dynamic transformation of HPL phase has not been investigated yet in detail. The HPL phase exhibits a double directionality of order-order phase transition when is subject to thermal process. Experimental evidences suggest that HPL phase of the PS-PI copolymer can evolve to a cylindrical structure and to a Gyroid structure by temperature effect (Park et al., 2005; You et al., 2007).

The double directionality of order-order phase transitions that exhibits the classical phase of PS-PI copolymer is a topic interesting to control the morphology and physical properties of polymeric materials. The order-order phase transitions can be investigated in more detail through mesoscopic simulations than through experimentation. Mesoscopic simulations are efficient methods to investigate the physical processes of soft matter and their interactions with chemical environments (Rodríguez-Hidalgo et al., 2011; Ramos-Rodriguez et al., 2010). Offer a particularly useful way of exploring the matter transfer process and to make predictions that may be of interest for understanding and elucidating complex process such as the double directionality of order-order phase transitions. In the mesoscopic simulations the atoms of each molecule are not directly represented, but they are grouped together into beads (coarse-grained models), where a springs force reproduce the typical nature of them, and therefore can exhibit a real physical behaviour in multicomponent systems. In this section, we explored the double directionality of phase transition of HPL structure of the PS-PI copolymer from a mesoscopic point of view by mesoscopic simulations, where the phase evolution stages and transient ordered states are analysed.

3.1 Model and simulation method

To explore the order-order phase transition of the HPL phase, we employed Dissipative Particle Dynamics (DPD) simulations. The original DPD method was introduced by Hoogerbrugge and Koelman and was later modified by Groot, R.D. (Hoogerbrugge & Koelman, 1992; Koelman & Hoogerbrugge, 1993; Groot & Warren, 1997; Groot & Madden, 1998, 1999). The DPD method allows the study of high-molecular-weight systems as the polymeric materials. The coarse-graine

$$\mathbf{f}_i = m_i \frac{d\mathbf{v}_i}{dt} \qquad (8)$$

where r_i, v_i, m_i and $\mathbf{f}_i$ are the position, velocity, mass, and force, respectively, of bead i. Dimensionless units are used in DPD simulations; usually, the mass of each bead is set to 1 DPD mass units, which results in an equation between the force acting on a bead and its acceleration. Each particle is subject to soft interactions with its neighbours via three forces: conservative (F_{ij}^C), dissipative (F_{ij}^D) and random (F_{ij}^R). The total force acting on particle i is:

$$\mathbf{f}_i = \sum_{j \neq i} \left(F_{ij}^C + F_{ij}^D + F_{ij}^R \right) \qquad (9)$$

The conservative force F_{ij}^C is a soft repulsive force that acts between particles i and j. The dissipative force $F_{ij}{}^D$ corresponds to a frictional force that depends on both the positions and relative velocities of the particles. The random $f\,v_i = \frac{r_i}{dt}$ orce $F_{ij}{}^R$ is a random interaction between a bead i and its neighbour bead j. All forces vanish beyond a cut-off radius r_c, which is usually chosen as the reduced unit of length, $r_c \equiv 1$. The $F_{ij}{}^D$ and $F_{ij}{}^R$ forces act as a thermostat that conserves momentum and gives the correct hydrodynamics at sufficiently large time and length scales. These forces are given by:

$$F_{ij}^{C} = \begin{bmatrix} a_{ij}(1 - r_{ij})\hat{r}_{ij} & (r_{ij} \leq 1) \\ 0 & (r_{ij} > 1) \end{bmatrix}$$

(10)

$$F_{ij}^{D} = \left[-\gamma \omega^{D}(r_{ij})(v_{ij}.\hat{r}_{ij})\hat{r}_{ij} \right]$$

(11)

$$F_{ij}^{R} = \left[\sigma \omega^{R}(r_{ij})\xi_{ij}\hat{r}_{ij} \right]$$

(12)

where $r_{ij} = |\vec{r}_i - \vec{r}_j|$, $\hat{r} = \vec{r}_{ij} / r_{ij}$, γ is the dissipation strength, σ is the noise strength, $\omega^{D}(r_{ij})$ and $\omega^{R}(r_{ij})$ are weight functions of F_{ij}^{D} and F_{ij}^{R} forces, respectively, and a_{ij} is the maximum repulsive force between particle i and j. For the DPD system to have a well-defined equilibrium state that obeys Boltzmann statistics, the equilibrium temperature is defined as $k_B T = \sigma^2/(2\gamma)$. This condition fixes the temperature of the system and relates with the two DPD parameters γ and σ ($k_B T$ is usually chosen as the reduced unity of energy). The parameter a_{ij} (henceforth referred to as the bead–bead repulsion parameter or simply as the DPD interaction parameter) depends on the underlying atomistic interactions and is related to the parameter χ through:

$$a_{ij} = a_{ii} + \frac{k_B T \chi_{ij}(T)}{0.306}$$

(13)

In this way, a connection exists between the molecular character of the coarse-grained model and the DPD parameter. The parameter a_{ij} is given in terms of $k_B T$ (DPD reduced units). Equation (6) implies that, if the species are compatible, $\chi_{ij} \approx 0$ and therefore, $a_{ij} = 25$. Established procedures for mapping between the DPD and physical scales and for choosing the system temperature are not yet available. We therefore use Flory–Huggins theory (through the $\chi(T)$ dependence) to introduce the temperature into the DPD simulations. To calculate the a_{ij} values of Equation (6), we use the Hildebrand relation:

$$\chi = \frac{V_m}{RT}(\delta_1 - \delta_2)^2$$

(14)

where V_m and δ are the mean molar volume and solubility parameter, respectively. In the DPD simulation, the dynamic behaviour of order-order phase transition of the HPL structure is followed by integration of the equations of motion of each species using a modified version of the Verlet algorithm (Verlet, 1967). The integration of the equations of motion for each particle generates a trajectory through the system's phase, from which thermodynamic observables may be constructed by suitable averaging. Based on this information, the order-order phase transition can be observed. In this algorithm, the forces are still updated once per integration, thus there is virtually no increase in computational cost.

3.2 Coarse-grained models and parameterization

The molecular structure of PS-PI copolymer was built by means of the polymer builder module of Accelrys (Accelrys, 2006). The polymeric molecule with linear architecture contains a total of 300 repetitive units in the main chain. The molecular weight of the PS-PI copolymer presented an interval of 20798–25842 g/mol. The diblock copolymer molecule was replaced by a coarse-grained model constituted by 30 beads, where each bead represents a statistical segment (characteristic ratio, ($C_n \approx 10$) (Soto-Figueroa, et al 2005)), see Figure 12. The Equation (15) was used to map the real structure of diblock copolymer to statistical model (mesoscopic model).

$$C_{SGD} = \frac{M_P}{M_m\,(SSL)} \tag{15}$$

where C_{SGD}, M_P, M_m and SSL are bead number with Gaussian distribution, the molar mass of the block copolymer, molar mass of a repeat unit and means statistical segment level (characteristic ratio (C_n), or persistence length (L_p) or statistical Kuhn segment length (a_k)) respectively (Soto-Figueroa et al., 2007).

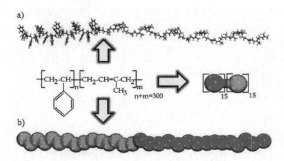

Fig. 12. Schematic representation of PS-PI copolymer: a) chemical structure of a PS-PI chain, n and m represent the polymeric unit number of each block, b) coarse-grained model of PS-PI system with linear architecture.

The chemical and physical nature of the coarse-grained model in mesoscopic simulation is described by interaction parameters (χ_{ij}). In order to study the order-order transition at temperatures different with DPD simulations we take the temperature dependence $\chi_{ij}(T) = \chi(T)$, in this way the real temperature is introduced into the DPD simulation (Rodríguez-Hidalgo et al., 2009). The interaction parameters for Equation (13) were evaluated from bulk atomistic simulations using the Fan, F.C. model (Fan et al., 1992). In this way the parameter interaction was expressed as:

$$\chi(T) = \frac{\Delta G(T)}{RT} = \frac{Z_{12}E_{12}(T) + Z_{21}E_{21}(T) - Z_{11}E_{11}(T) - Z_{22}E_{22}(T)}{2RT} \tag{16}$$

where ΔG denotes the Gibbs free energy, χ is interaction parameter, Z and ΔE_{12} are coordination number and differential energy of interaction of an unlike pair respectively. The Figure 13 show the interaction parameters $\chi(T)$, the temperature interval analysed is from 298 K to 500 K.

The tendencies that exhibit interaction parameters in temperature function are in agreement with the Hildebrand relation and are adequate to explore the formation of HPL phases (via

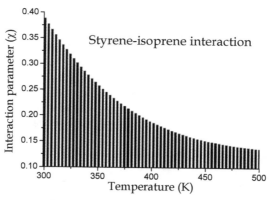

Fig. 13. Interaction parameters between styrene and isoprene molecules at temperatures different obtained by Monte Carlo molecular simulation (Vicente et al. 2006).

order-disorder transition) to room temperature and the double directionality of the HPL phase (via order-order transition) employing the coarse-grained model with predefined architecture proposed in this mesoscopic study.

All DPD simulations were performed in a cubic box that measured 20×20×20 in size, which contained a total of 2.4×10^4 representatives particles, a spring constant of C=4, and a density of ρ=3. The interaction parameters between identical species were then chosen as $a_{ST-ST} = a_{PI-PI}$= 25. Each bead was assigned a radius of 1. The coarse-grained number for each chemical species was held constant during the DPD simulations.

3.3 Simulation results and discussion

3.3.1 Hexagonally perforated layers structures

The equilibrium phases formation of PS-PI copolymer is governed by the composition, temperature and immiscibility between their components blocks, these factors were considered in the coarse-grained model via the architecture and interaction parameters.

All simulations start from a disordered state, where the PS-PI chains are in a homogeneous melted phase. First, we set the interactions parameters at T = 298 K. We then let the simulation proceed for 5×10^5 steps and during the temperature relaxation; we observed the microphase segregation process and the generation of equilibrium phases via ODT. Several transient stages were detected in the ODT process; a) melt phase, where the copolymer chains move freely; b) microphase segregation by temperature effect (temperature decrease), c) generation of pure microdomains of poly(styrene) and poly(isoprene), in this stage the ordered phases system attains an equilibrium temperature to room temperature.

To identify the composition region where the HPL structures are formed, we scanned the composition interval from 0.1–0.5 (volume fraction of poly(styrene) with increments of 0.03) at constant temperature of 298 K (i.e. constant χN). All DPD simulations generate a coarse-grained system sufficiently large to observe the classical phases formation in the analysed composition interval. The mesoscopic simulation of PS-PI copolymer exhibits a wide variety of structures or equilibrium phases such as LAM, HPC, BCC, Gyroid and HPL. The equilibrium phases of PS-PI system depend primarily on three factors: (i) the volume fraction of PS and PI blocks (f), (ii) the degree of polymerization (N), and (iii) the interaction parameter (χ). The results obtained by DPD simulations are in accordance with the mean-field theory (Leibler, 1980).

The HPL phase was obtained into the composition intervals of 0.3 to 0.36 (volume fraction of poly(styrene)), see Figure 14. The tendency of PS-PI chains to self-assemble into HPL structures depends of the previous factors and is governed by thermodynamics interactions (enthalpic and entropic) during the microphase segregation process.

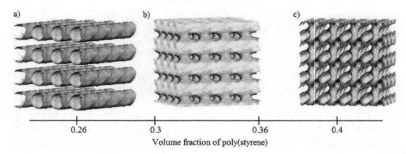

Fig. 14. Specific phases of PS-PI copolymer obtained as the blocks composition (PS/PI) is modified: a) cylindrical, b) HPL and c) Gyroid.

The thermodynamic stability of HPL phase in the predominance region vary with the relative chain length (poly(styrene) composition) of the component blocks. The predominance region where the HPL phase is formed has two composition limits that are contiguous with other equilibrium phases of different morphology. In the low composition limit where the volume fraction of poly(styrene) is close to 0.3, the HPL structure has as neighbour the cylindrical phase, whereas in the height composition limit (close to 0.36), the HPL structure has as neighbour the Gyroid structure. These behaviours between different phases are in accordance with the phase diagram reported by Khandpur (Khandpur et al., 1995).

The order-order phase transitions that exhibits the HPL structures with specific compositions of 0.3 and 0.36 (volume fraction of poly(styrene) are investigated. The HPL structures are modified through temperature increase, keeping constant the PS-PI block composition. The OOT of HPL structure for each specific composition shows a selective directionality (from HPL to cylinders and from HPL to Gyroid) during thermal study.

3.3.2 Order-order phase transition of HPL to cylindrical phase

The HPL phase with a specific composition of 0.3 (volume fraction of poly(styrene) was put at continuous cycles of thermal heating in the temperature interval from 298 to 500 K. A total of 2.0×10^5 time steps with step size of $\Delta t = 0.03$ were allowed in the mesoscopic simulation to reach the thermodynamic balance of PS-PI system in each temperature increment. The thermally induced phase transition from HPL to cylindrical phase was observed at the temperature of 432 K, this corresponds to the OOT, and this fact is in accordance with theoretical and experimental evidences (You et al., 2007).

The OOT process was visualized during mesoscopic simulation; the results show that the transformation from HPL to cylindrical phase by temperature effect is generated in several stages. When the HPL phase is annealed below the OOT temperature, we observed a slow dynamic motion of the PS and PI microdomains, the thermodynamic stability between microdomains different they are in an energetic barrier that maintain the HPL phase stable.

When the HPL phase is annealed to a higher temperature, $T \geq 432$ K, the energetic barrier that maintain the HPL phase stable is exceeded, the thermodynamic interactions (enthalpic and entropic) play an important role in the OOT process. The enthalpic interaction is proportional to the Flory-Huggins interaction parameter (Fig. 13), which is found to be inversely proportional to temperature.

During the temperature increase, the interaction parameter between the PS and PI microdomains diminishes, generating the OOT from HPL arrangement to cylindrical phase, the enthalpic interaction in the OOT is accompanied by an increase in entropy. At higher temperatures the entropic interactions dominates and is cause of anisotropic composition fluctuations into polymeric microdomains (Ryu, et al., 1999).

The HPL phase develops short-lived transient states during the OOT process, because of interface instability (PS/PI), combined with fast molecular motion of PS and PI microdomains. In Figure 15(a-d) are shown snapshots of the thermally induced phase transition from HPL to cylindrical structure during the annealed process at temperature, T= 432 K.

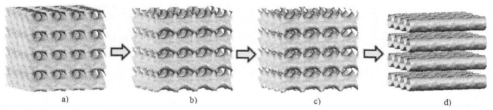

a)   b)   c)   d)

Fig. 15. Snapshots of the OOT pathway from HPL to cylindrical phase: a) undulation process of HPL phase, b-c) instability of HPL phase and d) formation of cylindrical phase.

The snapshot of Figure 15(a) corresponds to the initial equilibrium state of HPL phase, initially, the PS perforated microdomains display an undulation process induced by thermodynamic instability, in this point the PS perforated microdomains and the PS/PI interface become less rigid by temperature effect. The PS microdomains maintain their shape an short-period of time, as time goes, the anisotropic composition fluctuations (thermally-induced) increase quickly, the PI interconnections into PS perforated microdomains enlarge their area, in this stage the PS microdomains are unstable see Figure 15(b-c). When the anisotropic composition fluctuations reach a critical point, the PS unstable microdomains change their structure to a new cylindrical arrangement, see Figure 15(d), in this thermally induced stage the uniform cylindrical microdomains are stabilized.

3.3.3 Order-order phase transition of HPL to Gyroid phase

The HPL phase with a volume fraction of PS, $f_{PS}=0.36$, (obtained within the first 5×10^5 time steps) was now subject to a thermal heating process for another 5×10^5 time steps. The simulation outcome shows an OOT from HPL to Gyroid phase at the temperature of 438 K. The composition increase of poly(styrene) in HPL microdomains modify the phase transition directionality. The OOT from HPL to Gyroid phase obtained by DPD simulation are consistent with experimental results reported by Insun Park. They have investigated the phase transition behaviour from the hexagonally perforated layer (HPL) to the Gyroid phase in supported thin film of a poly(styrene)-*b*-poly(isoprene) (PS-*b*-PI) diblock copolymer (OOT occur at temperature of 443 K) (Park et al., 2005).

During the thermal heating process, the HPL phase undergoes the microdomains modification of PS and PI, generating transient intermediate stages as are sketched schematically in Figure 16(a-c). Three transient intermediate stages were detected during the OOT: (i) undulation process of pure microdomains and PS/PI interface due to the thermodynamic instability by temperature effect, see Figure 16(a), in this stage, the entropic and enthalpic interactions govern the microphase stability and induce anisotropic composition fluctuations into pure microdomains, (ii) increase of volume of poly(isoprene) interconnections into PS perforates microdomain, Figure 16(b), in this stage the PS perforated microdomains are instable, (iii) formation of parallel interconnections between PS perforated layers, see Figure 16(c), in this stage, the Gyroid phase is formed.

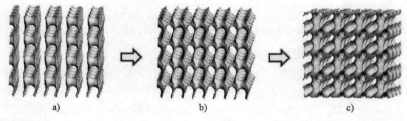

a)   b)   c)

Fig. 16. Snapshots of evolution process of the OOT from HPL to Gyroid phase: a) undulation process of PS microdomains in the PI matrix by anisotropic composition fluctuations effect, b) volume increase of PI interconnections into PS perforates microdomains, c) formation of Gyroid phase by generation of parallel interconnections between PS perforated microdomains.

The HPL phase of PS-PI diblock copolymer exhibits a double directionality of order-order transition from HPL to cylindrical phase and from HPL to Gyroid phase. The anisotropic composition fluctuations of PS and PI microdomains of HPL phase by thermal heating process induce the order-order phase transitions, however the phase transition directionality towards a specific phase (cylindrical or Gyroid) in the thermal process is governed by small variations of poly(styrene) concentration in the poly(styrene) microdomains of HPL phase. Variations of small composition into structured microdomains according to Leibler´s mean-field theory greatly modify the phase thermodynamic behaviour (separation, segregation and phase transformation).

4. Conclusion

The block copolymers are "smart" materials that have the ability to self-assemble inside a variety of periodic phases of high regularity in size and shape. The phase behaviour (type of structure and morphology) that exhibits these polymeric materials can be controlled by two different processes: order-disorder phase transition and thermotropic order-order phase transition. In the order-disorder phase transition, the phase behaviour is governed by entropic and enthalpic interactions, whereas the order-order phase transition are controlled by anisotropic composition fluctuations of theses thermodynamic processes can emerge supramolecular structures with well-defined morphologies and specific properties.

The ODT and OOT that exhibits of block copolymer can be explored in more detail through mesoscopic simulations than through experimentation. The mesoscopic simulation methods, offer a particularly useful way of exploring the phase behaviour pathway and to make predictions that may be of interest for understanding and elucidating complex process such as order-disorder of order-order phase transitions.

The DPD approach has been successfully applied to the investigation of phase transition processes (ODT and OOT) of HPL structure. The mesoscopic simulation outcomes show that the HPL phase of PS-PI diblock copolymer exhibits a double directionality of order-order phase transition from HPL to cylindrical phase and from HPL to Gyroid phase. This double directionality of OOT is controlled by small variations of poly(styrene) concentration in the poly(styrene) microdomains of HPL phase. Finally, all the simulation outcomes are qualitatively consistent with the experimental results, demonstrating that the DPD method may provide a powerful tool for the investigation and analysis of soft matter transformation process by thermal heating effects.

5. Acknowledgment

This work was supported by the Universidad Nacional Autónoma de México (UNAM), PAPIIT project No. IN104410-2 and IN109712. We also acknowledges the financial support of the Consejo Nacional de Ciencia y Tecnología (CONACYT) Project: 2012.

6. References

Almdal, K.; Koppi, K.A.; Bates, F.S.; Mortensen, K. (1992). Multiple ordered phases in a block copolymer melt. *Macromolecules*, Vol. 25(6), pp. 1743-1751.

Accelrys. (2006), Material studio release, notes, release 5.0; Accelrys Software, Inc.: San Diego, CA.

Bates, F.S. & Fredrickson, G.H. (1990). Block copolymer thermodynamics: Theory and experiment. *Annu. Rev. Phys. Chem.*, Vol. 41, pp. 525-557.

Bates, F.S. & Fredrickson, G.H. (1999), Block copolymers-Designer soft materials. *AIP Phys. Today*, 2, 32-38.

Bates, F. S. (1991). Polymer – Polymer phase behavior. *Science*, Vol. 251, pp. 898.

Bodycomb, J.; Yamaguchi, D.; Hashimoto, T. (2000). A Small-Angle X-ray Scattering Study of the Phase Behavior of Diblock Copolymer/Homopolymer Blends. *Macromolecules*, Vol. 33(14), pp. 5187-5197.

Balta-Calleja, F.J. & Roslaniec, Z. (2000), Block Copolymers; Marcel-Dekker Publishers, (Eds.): New York.

Court, F.; Yamaguchi, D.; Hashimoto, T. (2006). Morphological Studies of Binary Mixtures of Block Copolymers: Temperature Dependence of Cosurfactant Effects. *Macromolecules*, Vol. 39(7), pp. 2596-2605.

Fan, F.C.; Olafson, B.D.; Blanco, M. (1992). Application of molecular simulation to derive phase diagrams of binary mixtures. *Macromolecules*, Vol. 25, pp. 3667-3676.

Flory, P. J. (1953). Principles of Polymer Chemistry, Cornell University Press, Ithaca, New York,

Groot, R. D. & Warren, P. B. (1997). Dissipative particle dynamics: bridging the gap between atomistic and mesoscopic simulation. *J. Chem. Phys.*, Vol. 107, pp. 4423-4435.

Groot, R. D. & Madden, T. J. (1998). Dynamic simulation of diblock copolymer microphase separation. *J. Chem. Phys.*, Vol. 108, pp. 8713-8724.

Groot, R. D. & Madden, T. J., Tildesley, D. J. (1999). On the role of hydrodynamic interactions in block copolymer microphase separation. *J. Chem. Phys.*, Vol. 110, pp.9739-9749.

Hoogerbrugge, P. J. & Koelman, J. M. V. A. (1992). Simulating microscopic hydrodynamic phenomena with Dissipative particle dinamics. *Europhys. Lett.*,Vol. 19, pp.155-160.

Hajduk, D. A.; Harper, P.E.; Gruner, S. M.; Honeker, C. C.; Thomas, E. L.; Fetters, L. J. (1995). A Reevaluation of Bicontinuous Cubic Phases in Starblock Copolymers. *Macromolecules*, Vol. 28, pp. 2570-2573.

Hamley, I.W. (1998). The Physics of Block Copolymers. Oxford University Press, ISBN 019850218 4.

Koelman, J. M. V. A. & Hoogerbrugge, P. J.(1993). Dynamic simulation of hard sphere sispension under steady shear. *Europhys. Lett.*, Vol. 21, 363-368.

Kimishima, K.; Koga, T.; Hashimoto, T. (2000). Order−Order Phase Transition between Spherical and Cylindrical Microdomain Structures of Block Copolymer. I. Mechanism of the Transition. *Macromolecules*, Vol. 33(3), pp. 968-977.

Kim, E.Y.; Lee, D.J.; Kim, J.K. (2006). Phase Behavior of a Binary Mixture of a Block Copolymer with Lower Disorder-to-Order Transition and a Homopolymer. *Macromolecules*, Vol. 39(25), pp. 8747-8757.

Kim, J.K.; Lee, H.H.; Gu, Q.J.; Chang, T.; Jeong, Y.H. (1998). Determination of Order−Order and Order−Disorder Transition Temperatures of SIS Block Copolymers by Differential Scanning Calorimetry and Rheology. *Macromolecules*, Vol. 31(12), pp. 4045-4048.

Krishanamoorti, R.; Modi M.A.; Tse, M.F.; Wang H.-C. (2000). Pathway and Kinetics of Cylinder-to-Sphere Order−Order Transition in Block Copolymers. *Macromolecules*, Vol. 33, pp. 3810-3817.

Krishanamoorti, R.; Silva A. S.; Modi M.A. (2000). Small-Angle Neutron Scattering Study of a Cylinder to Sphere order-order Transition in Block Copolymers. *Macromolecules*, Vol. 33, pp. 3803-3809.

Khandpur, A.K.; Forster, S. ; Bates, F.S.; Hamley, I.W.; Ryan, A.J.; Bras, W.; Almadal, K.; Mortensen, K. (1995). Polyisoprene-Polystyrene Diblock Copolymer Phase Diagram near the Order-Disorder Transition. *Macromolecules*, Vol 28(26), pp. 8796-8806.

Leibler, L. (1980). Theory of Microphase Separation in Block Copolymers. *Macromolecules*, Vol. 13(6), pp. 1602-1607.

Matsen, M. W. & Bates, F. S. (1996). Unifying Weak- and Strong-Segregation Block Copolymer Theories. *Macromolecules*, Vol. 29, pp. 1091-1098.

Modi, M.A.; Krishanamoorti, R.; Tse, M.F.; Wang, H.C. (1999). Viscoelastic Characterization of an Order−Order Transition in a Mixture of Di- and Triblock Copolymers. *Macromolecules*, Vol. 32(12), pp. 4088-4097.

Mani, S.; Weiss, R. A.; Cantino, M. E.; Khairallah, L. H.; Hans, S. F.; Williams, C. E. (2000). Evidence for a Thermally Reversible Order-Order Transition between Lamellar and Perforated Lamellar Microphases in a Triblock Copolymer. *Eur. Polym. J.*, Vol. 36, pp. 215–219.

Park, I.; Lee, B.; Ryu, J.; Im, K.; Yoon, J.; Ree, M.; Chang, T. (2005). Epitaxial Phase Transition of Polystyrene-b-Polyisoprene from Hexagonally Perforated Layer to Gyroid Phase in Thin Film. *Macromolecules*, Vol. 38, pp.10532-10536.

Ramos-Rodriguez, D. A.; Rodriguez-Hidalgo, M. R.; Soto-Figueroa, C.; Vicente, L. (2010). Molecular and mesoscopic study of ionic liquids and their use as solvents of active agents released by polymeric vehicles. *Molecular Physics*, Vol. 108(5), pp. 657–665.

Rodríguez-Hidalgo, M.R.; Soto-Figueroa, C.; Vicente, L. (2011). Mesoscopic simulation of the drug release mechanism on the polymeric vehicle P(ST-DVB) in an acid environment. *Soft Matter*, Vol. 7, pp 8224-8230.

Rodríguez-Hidalgo, M.R.; Soto-Figueroa, C.; Vicente, L. (2009). Mesoscopic study of cylindrical phases of poly(styrene)-poly(isoprene) copolymer: Order-order phase transitions by temperature control. *Polymer*, Vol. 50, pp 4596-4601.

Ryu, C.Y. & Lodge, T.P. (1999). Thermodynamic Stability and Anisotropic Fluctuations in the Cylinder-to-Sphere Transition of a Block Copolymer. *Macromolecules*, Vol. 32, pp. 7190-7201.

Sakurai, S.; Kawada, H.; Hashimoto, T.; Fetters, L.J. (1993).Thermoreversible Morphology Transition between Spherical and Cylindrical microdomains of Block Copolymers. *Macromolecules*, Vol. 26(21), pp. 5796-5802.

Sakurai, S.; Hashimoto, T.; Fetters, L.J. (1996). Thermoreversible Cylinder–Sphere Transition of Polystyrene-block-polyisoprene Diblock Copolymers in Dioctyl Phthalate Solutions. *Macromolecules*, Vol. 29(2), pp. 740-747.

Sakamato, N.; Hashimoto, T.; Han, C.D.; Kim, D.; Vaidya, N.Y. (1997). Order–Order and Order–Disorder Transitions in a Polystyrene-block-Polyisoprene-block-Polystyrene Copolymer. *Macromolecules*, Vol. 30(6), pp. 1621-1632.

Soto-Figueroa, C.; Rodríguez-Hidalgo, M.R.; Martínez-Magadán, J.M. (2005). Molecular simulation of diblock copolymers; morphology and mechanical properties. *Polymer*, Vol. 46, pp. 7485-7493.

Soto-Figueroa, C.; Luis-Vicente.; Martínez-Magadán, J.M.; Rodríguez-Hidalgo, M.R. (2007). Self-Organization Process of Ordered Structures in Linear and Star Poly(styrene)-Poly(isoprene) Block Copolymers: Gaussian Models and Mesoscopic Parameters of Polymeric Systems. *J. Phys. Chem. B*, Vol. 111, pp. 11756-11764.

Soto-Figueroa, C.; Luis-Vicente.; Martínez-Magadán, J.M.; Rodríguez-Hidalgo, M.R. (2007). Mesoscopic simulation of asymmetric-copolymer/homopolymer blends: Microphase morphological modification by homopolymer chains solubilization. *Polymer*, Vol. 48, pp. 3902-3911.

Soto-Figueroa, C.; Rodríguez-Hidalgo, M.R.; Martínez-Magadán, J.M.; Luis-Vicente. (2008). Dissipative Particle Dynamics Study of Order-Order Phase Transition of BCC, HPC, OBDD, and LAM Structures of the Poly(styrene)-Poly(isoprene) Diblock Copolymer. *Macromolecules*, Vol. 41, pp. 3297-3304.

Soto-Figueroa, C.; Rodríguez-Hidalgo, M.R.; Martínez-Magadán, J.M.; Luis-Vicente. (2008). Mesoscopic simulation of metastable microphases in the order–order transition from gyroid-to-lamellar states of PS–PI diblock copolymer. *Chemical Physics Letters*, Vol. 460, pp. 507-511.

Strobl, G. (1997). The Physic of Polymers. Springer, ISBN 3-540-63203-4.

Thomas, E.L. & Lescanec, R.L. (1995). Phase morphology in block copolymer systems, In A. Keller, M. Warner and A.H. Windle (Eds.), Self-order and Form in Polymeric Materials, p.147-164. Chapman & Hall, London.

Verlet, L. (1967). Computer "Experiments" on Classical Fluids. I. Thermodynamical Properties of Lennard- Jones Molecules. *Phys. Rev.*, Vol.159, pp. 98-103.

Vicente, L.; Soto-Figueroa, C.; Pacheco-Sanchez, H.; Hernadez-Trujillo, J.; Martínez Magadán, J. M. (2006). Fluid Phase Equilibria. Vol. 239, pp. 100-106.

You, L. Y.; Chen, L.J.; Qian, H.J.; Lu, Z.Y. Microphase Transitions of Perforated Lamellae of Cyclic Diblock Copolymers under Steady Shear. *Macromolecules*, 2007, Vol. 40(14), pp. 5222-5227.

Permissions

The contributors of this book come from diverse backgrounds, making this book a truly international effort. This book will bring forth new frontiers with its revolutionizing research information and detailed analysis of the nascent developments around the world.

We would like to thank Dr. Zeeshan Nawaz and Prof. Dr. Shahid Naveed , for lending their expertise to make the book truly unique. They have played a crucial role in the development of this book. Without their invaluable contribution this book wouldn't have been possible. They have made vital efforts to compile up to date information on the varied aspects of this subject to make this book a valuable addition to the collection of many professionals and students.

This book was conceptualized with the vision of imparting up-to-date information and advanced data in this field. To ensure the same, a matchless editorial board was set up. Every individual on the board went through rigorous rounds of assessment to prove their worth. After which they invested a large part of their time researching and compiling the most relevant data for our readers. Conferences and sessions were held from time to time between the editorial board and the contributing authors to present the data in the most comprehensible form. The editorial team has worked tirelessly to provide valuable and valid information to help people across the globe.

Every chapter published in this book has been scrutinized by our experts. Their significance has been extensively debated. The topics covered herein carry significant findings which will fuel the growth of the discipline. They may even be implemented as practical applications or may be referred to as a beginning point for another development. Chapters in this book were first published by InTech; hereby published with permission under the Creative Commons Attribution License or equivalent.

The editorial board has been involved in producing this book since its inception. They have spent rigorous hours researching and exploring the diverse topics which have resulted in the successful publishing of this book. They have passed on their knowledge of decades through this book. To expedite this challenging task, the publisher supported the team at every step. A small team of assistant editors was also appointed to further simplify the editing procedure and attain best results for the readers.

Our editorial team has been hand-picked from every corner of the world. Their multi-ethnicity adds dynamic inputs to the discussions which result in innovative outcomes. These outcomes are then further discussed with the researchers and contributors who give their valuable feedback and opinion regarding the same. The feedback is then collaborated with the researches and they are edited in a comprehensive manner to aid the understanding of the subject.

Apart from the editorial board, the designing team has also invested a significant amount of their time in understanding the subject and creating the most relevant covers. They scrutinized every image to scout for the most suitable representation of the subject and create an appropriate cover for the book.

The publishing team has been involved in this book since its early stages. They were actively engaged in every process, be it collecting the data, connecting with the contributors or procuring relevant information. The team has been an ardent support to the editorial, designing and production team. Their endless efforts to recruit the best for this project, has resulted in the accomplishment of this book. They are a veteran in the field of academics and their pool of knowledge is as vast as their experience in printing. Their expertise and guidance has proved useful at every step. Their uncompromising quality standards have made this book an exceptional effort. Their encouragement from time to time has been an inspiration for everyone.

The publisher and the editorial board hope that this book will prove to be a valuable piece of knowledge for researchers, students, practitioners and scholars across the globe.

List of Contributors

A.M.S. Costa, F.C. Colman, P.R. Paraiso and L.M.M. Jorge
Universidade Estadual de Maringa, Brazil

Ramzan Naveed and Shahid Naveed
Department of Chemical Engineering, University of Engineering and Technology, Lahore, Pakistan

Zeeshan Nawaz
Chemical Technology Development, STCR, Saudi Basic Industries Corporation (SABIC), Kingdom of Saudi Arabia

Werner Witt
Lehrstuhl Anlagen und Sicherheitstechnik, Brandenburgicshe Technische Universität, Cottbus, Germany

Giorgio Rovero, Massimo Curti and Giuliano Cavaglià
Politecnico di Torino, K&E Srl, Italy

Jaime Alfonzo Irahola
Universidad Nacional de Jujuy, Argentina

Fábio de Ávila Rodrigues and Reginaldo Guirardello
State University of Campinas, School of Chemical Engineerging, Brazil

Khaled Elsaid and Ahmed Abdel-Wahab
Department of Chemical Engineering, Texas A&M University at Qatar, Education City, Doha, Qatar

Nasr Bensalah
Department of Chemical Engineering, Texas A&M University at Qatar, Education City, Doha, Qatar
Department of Chemistry, Faculty of Sciences of Gabes, University of Gabes, Gabes, Tunisia

Touhami Mokrani
University of South Africa, South Africa

Deresh Ramjugernath
Thermodynamic Research Unit, School of Chemical Engineering, University KwaZulu Natal, Howard College Campus, Durban, South Africa

Christophe Coquelet
Thermodynamic Research Unit, School of Chemical Engineering, University KwaZulu Natal, Howard College Campus, Durban, South Africa
MINES ParisTech, CEP/TEP - Centre Énergétique et Procédés, Fontainebleau, France

Maria Giovanna Buonomenna and Giovanni Golemme
Department of Chemical Engineering and Materials and INSTM Consortium, University of Calabria, Rende (CS), Italy

Enrico Perrotta
Department of Ecology, University of Calabria, Rende (CS), Italy

César Soto-Figueroa and María del Rosario Rodríguez-Hidalgo
Departamento de Ciencias Químicas, Facultad de Estudios Superiores Cuautitlán, Universidad Nacional Autónoma de México (UNAM), Mexico

Luis Vicente
Departamento de Física y Química Teórica, Facultad de Química Universidad Nacional Autónoma de México (UNAM), México